Climate Change Policy
in the European Community

Report of a workshop
held at
the Royal Institute of International Affairs
October 1992

Pier Vellinga and Michael Grubb (eds)

Energy and Environmental Programme
The Royal Institute of International Affairs
London

363.7387
C639

Produced as one of a series of Occasional Papers by the Energy and Environmental Programme.

Published 1993 by the Royal Institute of International Affairs, 10 St James's Square, London, SW1Y 4LE.

Exclusive distribution in the USA and Canada by The Brookings Institution, 1775 Massachusetts Avenue NW, Washington DC, 20036-2188.

ISBN 0 905031 66 0

MB

Research by the Energy and Environmental Programme is supported by generous contributions of finance and professional advice from the following organizations:

Amerada Hess • Arthur D Little • Ashland Oil • British Coal • British Nuclear Fuels • British Petroleum • Caltex • Chubu Electric Power Co • Commission of the European Communities • Department of Trade and Industry • Department of the Environment • Eastern Electricity • East Midlands Electricity • ELF(UK) • Enterprise Oil • Exxon • Idemitsu • Japan National Oil Corporation • Kuwait Petroleum • LASMO • Mitsubishi Research Institute • Mobil • National Grid • National Power • Neste • Nuclear Electric • Petronal • PowerGen • Shell • Statoil • Texaco • Tokyo Electric Power Co • Total • UK Atomic Energy Authority

Contents

Rapporteur's preface

In October 1992, the Royal Institute of International Affairs organized an international workshop on the climate policy of the European Community in the aftermath of the Rio Conference.[1] In the typical Chatham House tradition an often heated debate ensued with senior policymakers from the EC governments and the Commission, and others from business, academia and interest groups. All participants took part in the debate in their personal capacity and no record is kept of the individual statements. This report records the issues that were discussed, and the papers presented to the workshop.

An ambitious start

The development of climate policy at the European Commission level started off ambitiously in October 1990, just before the Second World Climate Conference, when the EC and its Member States decided to stabilize their emission of CO_2 at 1990 level. When, in the years following, it became clear that the USA as one of the important emitters of CO_2 did not follow this initiative, the EC enthusiasm decreased. Less than two years later the proposal for an EC-wide energy/carbon tax to support the stabilization policy failed to be adopted. The compromise in the EC was that introduction of such a tax would wait until similar steps were taken by other major economic actors such as the United States and Japan; but the EC reaffirmed its commitment to the stabilization target.

At the Rio Conference all OECD countries, including the Member States and the EC as a whole, agreed to take actions individually or jointly with the aim to return the level of CO_2 emissions to 1990 levels by the year 2000. Although this is not a legally binding commitment, the countries agreed to present national programmes that reflect the measures taken to achieve such an aim. Already the USA is intensifying its policies to limit the emission of CO_2 and President Clinton in his 'State of the Union Address' on February 17th 1993 announced plans for an energy tax.

This puts an increasing pressure on the EC to fulfil its commitments. Many of the Member States are taking measures to limit the emission of CO_2. But in the open market there are limits to what a Member State can do. The EC Commission is hesitant to come forward with proposals as the previous proposals were either crippled by the Member States' amendments, like the SAVE programme, or disapproved of like the tax proposal.

---

1. The United Nations Conference on Environment and Development, held at Rio de Janeiro in June 1992.

The pressure to take action will mount as the EC and the Member States ratify the Climate Convention signed at Rio.[2] Following ratification, the EC and its Member States will have to submit reports to the Conference of Parties which show *inter alia* that they are aiming to return CO_2 emissions in the EC as a whole to 1990 levels by the year 2000.

The EC climate policy dilemma

The EC CO_2 stabilization target was decided upon in October 1990 with the clear understanding that countries like Spain, Portugal, Greece, and Ireland would be allowed to increase their relatively low CO_2 emissions. The present situation puts these countries in a difficult position. They cannot ratify the Convention unless they can prove that other countries, or the EC as a whole, compensates for their growing emissions with reductions elsewhere. Spain has indicated that it can only support ratification by the EC when there is a full EC wide agreement on CO_2 burden sharing. This puts the EC Commission in an almost impossible position as some of the countries with growing emissions have fiercely resisted any of the EC proposals to introduce EC wide measures for CO_2 stabilization.

This is 'the EC climate policy dilemma'. At the workshop there was much debate over the priority of the EC-wide, versus the direct national obligations. In the report you will find a number of suggestions. A possible scenario is that the EC would reconsider the tax proposal after the USA has taken certain (albeit limited) steps of such a nature.

Taxes or standards?

At the workshop traditional and emotional views were exchanged about the pros and cons of an energy/carbon tax. One particular argument deserving attention here is the perception of energy intensive industries. They claim that a carbon tax would dramatically affect their global competitiveness. However, the governmental and the EC policymakers stress that compensation and/or exemption measures are envisaged to avoid any negative impacts. I believe that a renewed debate and policy preparation through rational discussions with the Member States and with industry may well provide practical solutions to this problem.

If the EC-wide tax proposal cannot be revived, then the only way to proceed is through greater regulation and the introduction of energy efficiency standards. In the workshop report you will find arguments regarding the introduction of this type of policy in relation to the concept of subsidiarity. What were the arguments used to weaken the SAVE proposals on efficiency standards? Are they valid? What is the role of energy pricing in relation to standard setting and research and development? The workshop participants were widely in agreement about the role of energy pricing as an instrument to encourage greater energy

---

2. *The UN Framework Convention on Climate Change,* discussed later in this report.

efficiency over the longer term. However, its short term effects and implications were heavily debated. For the business sector, energy policy development looks like having to choose between the devil (taxes) and Beelzebub (standards). However, for the entire group of workshop participants it was clear that the efficiency of energy use should be improved and that this does indeed require policy measures. A combination of taxes and standards in a sectoral specific mix is likely to be the best approach.

Challenging times ahead
While the debate on energy policies goes on, greenhouse gas concentrations continue to build up. Some may be happy that the eruption of Pinatubo provides a respite in the last decade's trend of higher global temperatures.[3] Others may be happy that the public response to the 'Maastricht Treaty' delays effective EC action. Meanwhile, the signals from the insurance industry indicate that the damage caused by extreme weather events is increasing. The pressure within and outside the EC to ratify the Climate Convention is increasing. Adoption of EC-wide policies, with appropriate agreement, is urgent.

Challenging and interesting times are ahead of us. The debate as reflected in this report provides the best up-to-date overview of the state of the art and the possible directions for further EC climate policy. The report provides not only opposing views, but is also a rich source of information on the development of and prospects for climate policy in the EC and its Member States.

Prof. dr. Pier Vellinga
Director, Institut voor Miliuvraagstukken (IVM)
The Netherlands

March 1993

---

3. The eruption of Mt Pinatubo in 1991 ejected more dust and aerosols into the upper atmosphere than any other eruption this century. Climate models indicate that this should cool the surface significantly over the subsequent few years, and some cooling has since been observed.

3

Part I: Background and workshop discussions

Introduction: EC global warming strategy and projections
Michael J. Grubb

As an introduction to this report a general overview is presented of the formal data, projections and emission targets as presented by the EC Member States and the Commission of the European Communities (CEC).

The European Community has taken a forward stance on the issue of climate change, and CO_2 emissions in particular. EC member countries were amongst the first to adopt targets for limiting CO_2 emissions, and to urge the international community to negotiate a binding Convention including emission constraints.

The decision with the single greatest impact on the development of the issue, both within the EC and on the broader international discussions, was the declaration by the joint Council of EC Energy and Environment Ministers of Member States, on 29 October 1990, that:

> The European Community and Member States assume that other leading countries undertake commitments along [similar] lines and, acknowledging the targets identified by a number of Member States ... are willing to take actions aiming at reaching stabilisation of the total CO_2 emissions by 2000 at 1990 level in the Community as a whole.

This declaration 'assumes' that other countries take measures, but does not make the goal explicitly conditional upon such action. In fact, all OECD countries except the US and Turkey did adopt some kind of stabilization goal, moves which were noted and welcomed in the Ministerial Declaration of the Second World Climate Conference. Reaffirmations by the Council of the EC aim expressed in the above declaration, statements by the EC in the process of negotiating the UN Climate Convention, and the results of those negotiations, have all increased the weight accorded to this commitment.

The goal was not a random choice. Nor did it reflect simply a recognition that such 1990-2000 stabilization[4] was fast becoming a standard symbolic and psychological demonstration to the developing world that developed countries could, and intended to, start addressing the problem by at least ensuring that their CO_2 emissions did not continue to rise. It was also a reflection that the targets already declared by member countries, if achieved, would be almost sufficient to achieve 1990-2000 stabilization across the EC.

---

4. The term 'stabilization', strictly interpreted, means that emissions should be capped permanently at current levels. In this report we use the term '1990-2000 stabilisation' to refer to the goal, as reflected in many unilateral emission targets and the declared aim for industrialised countries in the Climate Convention, that emissions in the year 2000 should not exceed the 1990 levels.

With the exception of the UK moving its target date for returning emissions to 1990 levels forward from 2005 to 2000, little has changed in the EC emissions outlook since 1990. Table 1 shows various measures of 1990 emissions by member countries, and the declared national emission targets. Set against these are the emissions projected in the absence of any abatement measures for the Reference Scenario recently calculated by the CEC's Energy Directorate.[5]

The table illustrates several points about the EC situation. Total emissions are dominated by Germany and the UK, and then Italy and France. There are wide differences in the starting per capita emissions, with those from Germany and the UK being about twice the level in Spain, with Portugal even lower. This reflects different patterns of economic development, but also other factors like climate and energy supply mix, as indicated by the relatively low per-capita level of France and Italy.

The reference projections indicate that CO_2 emissions, excluding the former East German territory ('old-EC'), are expected to rise about 13% above 1990 levels by 2000 in the absence of abatement measures. The Commission acknowledges considerable uncertainty in such projections - and outlines also a 'higher growth' scenario in which emissions by the year 2000 from the big four (Germany, UK, Italy and France) are 3-5% higher than in the reference case.[6]

However, the national targets set would keep old-EC emissions growth to 3.6% above 1990 levels, with nearly all the difference (compared with the CEC reference scenario) coming from abatement in Germany, Italy, the UK, Belgium and Denmark;[7] Spanish emissions would increase to nearly 10% of the total.[8] However, if the collapse in CO_2 emissions from the former East Germany projected by the Commission does materialize and is incorporated *and* the rest of Germany were to achieve the separate reductions illustrated in Table 1, the effect is large enough to make a big difference: if all declared emission targets were then kept to, then the assumptions of Table 1 indicate the 'new-EC' emissions in 2000 would be only 1.3% above the 1990 levels. Germany has confirmed that its objective of reducing CO_2 emissions by 25-30% by the year 2005 applies to the whole of Germany including the Five New Länder.

5. 'A View to the Future', *Energy in Europe*, Special Issue September 1992, CEC-DGXVII, Brussels.

6. ibid.

7. There is little additional saving from the Netherlands shown in the table because the Dutch government has already adopted a substantive National Environmental Policy Plan to limit emissions, so that abatement measures are already reflected in the Commission's reference scenario.

8. The CEC estimates in Table 1 may be compared with the figures presented to the workshop in the Spanish presentation, reproduced in the Spanish paper in Part II below, which shows much higher reference emissions and slightly higher 1990 and 2000 target emissions.

Table 1. EC CO_2 emissions: 1990 levels and projections for 2000

Country	CO_2 target emissions (target year, base year)	1990 emissions ($MtCO_2$)			Projected emissions in 2000 ($MtCO_2$)	
		% EC	Per capita	Total	CEC reference[1]	National targets
Belgium	-5% (2000,1990)	4.0%	11.2	112[1]	121.7	106
Denmark	-20% (2005,1988)	1.8%	9.9	51[1]	65.5	48[2]
France	stabilisation at 2tC/cap	13.2%	6.5	366[1]	431.4	425
Germany (West)	-25% (2005,1987)	25.5%	11.3	709	800.6	674[2]
Greece	+25% (2000,1990)	2.7%	7.4	74[1]	96.6	92
Italy	0% (2000,1990)	14.4%	6.9	400	464	400
Ireland	+20% (2000,1990)	1.1%	8.8	31	36	37
Luxembourg	0% (2000,1990)	0.5%	35.1	13[1]	13.7	13
Netherlands	-3% to -5% (2000,1989)	6.6%	12.2	182	178.1	177
Portugal	+29 to 39% (2000,1990)	1.4%	4.1	40	57	55
Spain	+25% (2000,1990)	7.6%	5.4	211[1]	259.8	263
UK	0% (2000,1990)	21.1%	10.2	587	614.1	587
EC12 % increase	0% (2000,1990)			2776	3138.5 13.1%	2877 3.6%
Germany (East)	16.23	10.7%	18.3	298	236.8	n.a.

7

Footnotes to table:

[1] Data from *A View to the Future*, Energy in Europe, Special Issue September 1992, CEC-DGXVII, Brussels. Otherwise data are taken from national plans or statements.
[2] Figures for countries which have a year 2005 target 20% reduction are estimated as a 5% reduction achieved by the year 2000, because many measures can contribute substantially only after 2000.

Such figures indicate the background situation and geographical spread of and trend in emissions and declared targets. But explicit target-sharing is not the route which the EC chose to take. Instead, in keeping with hopes for a more united EC, the European Commission was in 1990 asked by member governments to prepare an EC-wide strategy for turning the projections of 12-14% emissions increase into a collective stabilization. Extended discussions, and to some extent commandeering of existing EC programmes for addressing Europe's energy needs, led to a 5-point strategy being advanced by the Commission, backed up by a series of analyses and discussion documents:[9]

i) Direct measures to improve energy efficiency through implementation of the existing SAVE proposals[10] for a series of Directives on energy efficiency standards. It was estimated that these would reduce EC CO_2 emissions in 2000 by 3% below the reference projection.

ii) Strengthening of existing measures for promoting the dissemination of better energy conversion and use technologies, primarily through new phases in the EC's THERMIE programme. These were estimated to save another 1.5% of projected CO_2 emissions. The JOULE programme for energy RD&D would also encourage development of better technologies primarily for longer-term reductions.

---

9. 'A Community strategy to limit Carbon Dioxide emissions and to improve energy efficiency', SEC(91) 1744 final, Brussels, 14 October 1991; 'Draft communication on Community CO_2 stabilisation by the year 2000 - energy evaluation,' COM(92) 158/2, Brussels, 23 April 1992; 'A Community strategy to limit Carbon Dioxide emissions and to improve energy efficiency', COM(92) 246 final, Brussels, 1 June 1992; 'Proposal for a Council Directive introducing a tax on Carbon Dioxide emissions and energy', COM(92) 226 final, 30 June 1992.

10. Specific Actions for Vigorous Energy Efficiency: see the contribution in Part II by Andrew Warren.

iii) A programme of support for renewable energy technologies, which emerged as the proposed ALTENER Directive which set goals for the contribution of renewable energy. This would have most impact after the year 2000, but was projected to reduce emissions by another 1% by that year.

iv) A combined energy/carbon tax, to be introduced at a level of three dollars per barrel of oil equivalent ($3/boe) in 1993, rising by $1/boe annually to a level of $10/boe in 2000.[11] The reduction in year 2000 emissions would be 'between slightly more than 3% and some 5.5% of the 1990 level according to the policy stance on industry exemptions and the way of taxing electricity'.

v) Additional measures taken by Member States, which would report their national strategies to the Commission, which would then be empowered to monitor and review them, and if progress towards the target were inadequate, propose new measures.

Full implementation of the Community-wide measures would thus reduce projected emissions by 8.5-11%, leaving a rather small gap to be filled by additional Member State initiatives under the 'monitoring' proposals.

It was originally intended by the Commission (and indeed governments) that this package of measures would be agreed by the European Council of Ministers before the Rio 'Earth Summit' conference. However, agreement of a draft directive on the carbon/energy tax by the Commission proved to be very difficult. Concerns about the impact of the tax on industrial competitiveness led to substantial exemptions for energy-intensive industries, and it was decided (due partly to electricity trade complications) that the tax should apply to electricity output rather than input fuels; both weaken the impact on emissions, limiting the likely emission reductions to little more than 3% of the reference projection by 2000.[12] In a further crucial change, the tax proposals were also made:[13]

---

11. This was later clarified as a tax, split 50:50 on the energy and carbon components, starting in 1993 at 0.21 ECU/GJ (European Currency Units per gigajoule) plus 2.81 ECU/tCO$_2$, and rising by a third of this starting amount annually to a level of 0.7 ECU/GJ plus 9.4 ECU/tCO$_2$ in the year 2000. 1ECU = US$1.2 in late 1992. The final tax level equates to a total of around $80 per tonne of carbon. Small hydro and new renewable energy sources were exempted.

12. Modelling studies suggest that applying the tax to electricity output instead of fuels for power production reduces its impact on emissions by nearly a third (1.5 percentage points). The direct effect of industrial exemption is not so large, especially as the companies would in return be expected to take equivalent energy-saving measures.

13. COM(92) 226 final. Brussels, 30 June 1992.

'conditional on the introduction by other member countries of the OECD of a similar tax or of measures having an financial impact equivalent to the measures provided for in this Directive'.

Furthermore the SAVE programme was substantially weakened, with implementation of most measures being left to discretion of the Member States. At present it seems likely that implementation will at best be rather late and patchy. Thus, two central pillars of the EC strategy - coordinated standards to improve energy efficiency and to provide fiscal incentives - both appear greatly weakened.

Where does this leave EC climate strategy? As discussed in the papers to this workshop, the UN Framework Convention on Climate Change has reinforced the overall goal of stabilizing EC emissions, and added a requirement that the EC spell out its strategy for achieving this to international inspection. Yet agreement on the strategy proposed by the Commission, at the request of EC governments two years ago, does not now seem possible; the EC does not have an agreed approach to its collective goal.[14] What then are the prospects for EC climate policy? This is the background to this workshop.

---

14. See the Postscript to this report.

Report of workshop discussions
Matthew Paterson and Michael J. Grubb

Session 1 - EC policy and the Climate Change Convention

The primary purpose of this session was to examine the implications of the Framework Convention on Climate Change[15] on the EC's climate policy. It introduced some of the major issues for EC climate policy, such as:
- commitments in the convention and the implications for the EC;
- the CO_2/energy tax proposal;
- the division of responsibility between the EC and the Member States.

The opening presentation emphasized that the Convention is very important for understanding the obligations which the EC is now under. The EC target set in October 1990 was primarily a political agreement designed to influence the outcome of the Second World Climate Conference in November 1990, and subsequent negotiations on the Climate Convention. The importance of the Convention for the EC is that it turned that political agreement into a legally binding commitment 'to enact policies designed to limit emissions, with the aim of returning emissions to their 1990 levels by 2000'. It was suggested that the Convention provides a strong framework - states are obliged to draw up National Strategies concerning their emissions, and send them to the Secretariat.

An EC participant noted that the member countries of the EC had agreed to submit national plans to the Commission much earlier, as part of the EC strategy, but that by the target date (June 1992) the EC had received only some '2.5 plans and 6.5 letters' from the Member States. Such a level of response will be wholly insufficient in the context of the Convention obligations. The national plans, as now committed to, should be detailed and provide information by which it will be possible to judge states' actions and results of those actions.

Burden sharing/division of responsibilities
An ambiguity in the wording of the Convention was quickly brought up in relation to the EC's commitments. There remains a question about where ultimate responsibility for implementing the provisions of the Convention lie within the EC, since the EC is itself a signatory alongside the Member States.

---

15. *The UN Framework Convention on Climate Change*, signed at the UN Conference on Environment and Development, Rio de Janeiro, June 3-14 1993 by 153 states plus the EC. All member countries of the EC signed, and the Convention contains language specifically for regional bodies - the EC - which signed as a separate legal entity. The Convention will enter into force after 50 ratifications. A discussion of the Convention with key extracts is given in M.Grubb et al, *The Earth Summit Agreements: A Guide and Assessment*, RIIA/Earthscan, London / Brookings, Washington DC, April 1993.

The ambiguity is over the term 'individually or jointly' applied to the commitment to enact policies to return emissions to their 1990 levels by 2000. It was unclear to some participants whether this implied that all EC states were obliged to enact such policies. This would be a problem since some of the less developed Member States (for example Spain) had explicitly rejected the notion that they would agree to stabilize their own emissions in view of their low starting point. However, it was generally agreed that this section of the Convention meant that industrialized country signatories would be individually responsible unless they were involved in a joint agreement (such as the EC target). Spain was thus not obliged under the Convention to try to stabilize its emissions. But Spain and others in a similar position are obliged to demonstrate that they are indeed involved in a joint agreement which will achieve collective 1990-2000 stabilization[16] of CO_2 emissions.

There does however remain an ambiguous relationship between the EC's commitment and the Member States' commitments. This is primarily about the division of effort; what actions are to be taken at the EC level and what at the Member State level? Some EC action is clearly needed to help the states within the Community with relatively low but rapidly growing emissions (such as Spain, Greece, Portugal and Ireland) to implement policies. This is particularly so since the EC has not negotiated national quotas to meet the EC's commitment collectively. Another problem mentioned in this regard was the process by which the EC would negotiate how its commitment was going to be reached by the Member States collectively.

One suggestion was that the EC should take no further action until the Member States have drawn up and implemented their National Strategies; if the impact on emissions is then observed to be inadequate, further measures could then be developed. This was rejected by others as unacceptable as it would in effect leave the EC without a coordinated climate policy until about 1998, which would be far too late to meet the Convention commitments concerning the reporting of strategies and projections. Most participants accepted that the Community needs to have a coherent plan, with projections that are consistent with Member States' plans and which achieves the collective 1990-2000 stabilization objective, in time to present to the UN as agreed in the Climate Convention.

EC tax proposal
Discussion of the EC carbon/energy tax proposal in this session focused primarily upon the underlying rationales and international aspects; more detailed discussion of the impacts and implementation issues were considered in a later session summarized below.

---

16. See footnote 4.

Participants queried the meaning of making the tax conditional on 'equivalent action' by other OECD countries (see introductions). It was asked whether any specific set of possible measures which could come under this category is envisaged. For example, tradeable permits enacted in another country might reduce emissions but would not necessarily generate fiscal transfers similar to a tax - would this count as equivalent? One response was that tradeable permits can be used to raise revenue, depending on the system introduced. Others suggested that what mattered was the implementation of measures with similar impacts on costs. Ultimately, the decision about whether or not measures outside the EC were equivalent would be a political judgement by the EC Council.

Another participant argued that the market philosophy behind the Commission's CO_2 tax proposal is misleading. The market philosophy means that the market determines price. Only if the level of tax balances the costs of CO_2 abatement with those of carbon absorption does the level of the tax reflect a market-based philosophy. Absorbing carbon by planting trees may initially cost the equivalent of only $1-3/boe, well below the proposed level of carbon/energy tax of $10/boe.

A point offered in favour of an EC-wide tax is that while it is important to share the burden, it is very difficult to divide explicit emission targets. Quotas or tradeable permits may be too difficult to negotiate, in which case fiscal measures, such as the carbon tax, may be more appropriate.

Another participant noted that the EC should not be naive about the possible response of OPEC to a CO_2 tax - about the impact of carbon taxes on oil exporting countries, and their likely efforts to retaliate. Negotiations should be considered between OPEC and the EC on these issues.

Some of the objections to the EC measures, and the carbon tax proposals in particular, came from a general scepticism towards climate change as a problem requiring policies to limit EC emissions. It was suggested that significant sections of the scientific community still have doubts about the underlying science of climate change. Furthermore, the EC accounts for only 13-15% of total global CO_2 emissions. Thus EC stabilization is irrelevant in global terms. One participant characterized the proposals as a measure which would cost individuals £200/yr to fail to prevent a problem which may or may not arise 60-70 years in the future. Only a focus on a 'no cost' strategy is therefore justifiable in the EC.

However, this viewpoint was not widely accepted. Some of the reasons given for rejecting it were that it is known that human activity is changing the radiative properties of the atmosphere and that this is almost certain to disturb the climatic system. Scientists recognize the uncertainties; but the vast majority express great concern about the implications, and regard the climate models as the best indications we have as to the consequences. Also, the climatic system has very long time lags, so it is not possible to wait for direct observation and detailed understanding of the impacts because it would then be too late to avoid the consequences. Waiting is therefore very risky and undesirable. Furthermore, energy systems also have long time lags, and the industrialized countries as a whole are predominant in global emissions; the Convention reinforces the need for these countries to 'lead the way' if developing countries are to be involved at all. Such a policy would have major long-term technical and economic advantages as well. It was therefore suggested that industrialized countries stabilizing their emissions is a necessary first step.

Session 2 - National strategies and perspectives

This session was stimulated by presentations on three important and very different countries in the EC: the UK, Germany and Spain; these presentations are summarized in the papers in Part II of this report. The subsequent debate focused on three areas. First was how opinion has evolved on the climate issue within the Member States. Second was about policy aspects - in particular, what are the relative benefits of market-based and standards-based regulation? Third, what is the relevant distribution of responsibility between the Member States and the EC as a whole?

Questions about public, industrial and political perceptions on the greenhouse issue were introduced with reference to the Netherlands. There, the fact that at Rio and at the EC level much less was achieved than called for and expected has backfired in the national debate. As energy prices have declined, CO_2 emissions are growing more rapidly than government initiatives can cope with. As a result the national stabilization target is under pressure.

The situation in the UK regarding climate change is similar to other environmental issues. There is much support for action, until the distribution of costs is discussed. Public opinion is also much affected by the weather. The government is now in the position of having to lead the public to meet the national target. Another participant suggested that despite a low level of public understanding of the details of climate change, there is a general feeling in the UK that the government is not doing enough.

As regards Germany, it was suggested that global warming is still high on the agenda. After Rio, the pressure has not declined. The German Chancellor's initiative for the first meeting of the Conference of the Parties to take place in Germany is important. Parliament is debating the measures that need to be taken. Different economic sectors are involved - the presenter said that no sector has denied that the target of reducing emissions by 25% (by the year 2005) can be reached. All branches of industry see market openings for more efficient technologies. However, there are some problems, for example with the coal industry. The transport sector is also problematic. However, the debate there is driven by issues of urban congestion and other air pollution (NO_x). The need for investment in public transport is recognized by the Ministry of Transport.

Public opinion in Spain is also very concerned. This is one reason why the National Climate Commission was created. It is not made up simply of politicians. It has also scientists, industry and NGOs represented on it, and was created primarily to follow up the Rio conference.

For Denmark, the assessment was that the pressure is the same after Rio as previously. There is great political awareness. In mid-November 1992, the Danish government reported on Danish initiatives, with regard both to the Convention, and to Agenda 21.[17] There is already a carbon tax in Denmark, and the prospect of one in the EC is viewed as very desirable. The participant stated that Denmark intended to make this a major thrust of policy during the Danish presidency of the EC (January - June 1993). The Danes see this economic instrument as a prerequisite to energy efficiency and conservation.

In France, the situation is ambiguous. There is already both high energy and high CO_2 efficiency. It will therefore be difficult to stabilize emissions. There is a High-level Ministerial Group which is addressing French strategy. There have been demands for a 'no-cost' strategy. There are attempts to move Electricité de France more in the direction of renewables and energy efficiency. In the transport sector it will not be possible to stabilize emissions. A carbon tax will probably be necessary; the participant noted but did not share concerns about its impact on French industry.

Policy instruments, financial incentives and/or standards

Regarding policy options, many participants argued that the dichotomy between markets and regulation is a red herring. In different sectors, different measures will be needed. For example, transport and appliances are very amenable to standards, whereas other areas are not. Market mechanisms are useful both in addition to standards, and in areas where standard setting is not useful. One participant suggested that in the transport sector, car travel is very price-insensitive. Additional measures may therefore required, since congestion is little affected by carbon taxes.

Another asserted that basic standards are necessary within the EC. The alternative is that German standards will be universalized within the EC anyway, because of German economic strength. A basic set of standards throughout the EC would equalize the situation. The level of German standards were not objected to, but the level of EC standards should be decided in Brussels. However, the same participant also argued that use of the pricing mechanism is crucial.

---

17. Agenda 21 is the general programme of measures towards sustainable development agreed at the Rio Conference. A summary and analysis of this also is given in Grubb et al, *The Earth Summit Agreements*, op.cit. ref 14.

Division of responsibilities

On the issue of the relative roles of Member States and the EC as a whole, it was argued that some appliance regulation has to be at the EC level, for example vehicle and appliance regulation since these commodities are traded heavily in the EC. The harmonization of standards, especially in the area of transport, is therefore especially important. The argument was put forward that carbon/energy taxes should be organized at the EC level because differing tax rates would affect the relative competitiveness of Member States' economies. It was countered that this is not essential - countries differ widely in their current tax regimes and some have already introduced carbon taxes unilaterally. Other issues such as the detail of standards on non-traded goods, for example building regulations, should be addressed at the national level.

Session 3 - Industry and transport: impacts, options and opportunitiess

Much of the debate in this session focused upon carbon taxes, both in conceptual terms and in the form proposed by the EC Commission. There was no significant consensus as to whether a carbon tax would be an effective mechanism for reducing CO_2 emissions. In general, participants from industry expressed the view that a carbon tax would be too blunt an instrument, and could hit the competitiveness of European industry significantly. On the other hand, several participants expressed strongly the view that, at least in the long term, using a tax to increase the price of carbon fuels is essential in limiting long-term CO_2 emissions.

The arguments given against a carbon tax were the following. It was suggested that as a mechanism, taxing carbon is too blunt. It affects all carbon-using processes irrespective of how price sensitive the demand is in that area. One participant suggested that there should be simply be a tax on petrol. Another simply denied that price needs to be increased in order to reduce energy demand. A further argument was that a tax, while it might be designed to increase investment in energy conservation and efficiency, would decrease incentives to invest in general since it would reduce profits and be inflationary, and this would deter investment, for example in insulation.

Many of these points were contested by other participants. The debate focused in particular on the role of prices in influencing energy demand. Many participants suggested that in fact the role of price was crucial. A representative of one of the Member States stated that their conservation programmes had been less effective than anticipated because of the collapse in energy prices in the mid-1980s. Others suggested that for example, in France, domestic consumption for heating is low precisely because it is largely electric and electricity is relatively expensive. Another example was the size of cars in the USA, which historically can be attributed to low petrol prices. It was argued that while at the individual level some people clearly do respond to price changes but others clearly do not, in aggregate, price is very important in influencing long-run trends in consumption. It is important not simply in directly affecting the behaviour of firms and individuals, but in affecting the whole context within which policy decisions and planning take place. Most participants emphasized that the effect of price is cumulative over a long period, and that since climate change and energy investments are long term issues, tax changes to increase the price of energy need to be phased in early.

Another criticism made of those objecting to the EC's proposed carbon tax was based on their assessment of the costs of implementing it. It was argued that the EC's tax would increase a typical chemical company's energy bill by 25%. Since the EC's proposal is not only conditional but explicitly exempts such energy intensive industries, the relevance of this observation was challenged. So was its accuracy even if there were no industrial exemptions, since the assessment is based purely on the direct costs of the tax without either allowing for gains made through cuts in other taxes, or for the alleviating effects of investments that companies might make to reduce their energy consumption.

One participant claimed that the opposition of the chemical companies and other European industrial giants to the CO_2 tax (it was claimed that they had organized the largest-ever lobby of the Community by European industry against it) was disingenuous, in that they were professing to prefer voluntary commitments, but were not living up to those they had made. Another expressed surprise that attitudes had changed quickly in the business community, from being strongly in favour of market-based mechanisms only four years ago, to being opposed to them now. And another expressed concern that as industry had begun to express opposition to plans for a carbon tax, so had the public, since they did not see why they should be subjected to a tax while industry was not.

A distinction was drawn between a 'fundraiser' tax, intended primarily to raise money directly for investment for specific purposes (such as improving energy efficiency or R&D), and a tax designed primarily to affect behaviour through price signals. Some participants argued that the former would be preferable, and opposed the idea of 'revenue neutrality' in the EC proposals, which state that the tax should be offset against other reductions so as not to increase the overall tax burden. Some proponents of this recognized that the proposed EC tax of $10/boe might raise far too much money to make 'hypothecating' the revenues to particular uses either feasible or desirable, and suggested that a smaller hypothecated tax (perhaps raised directly by the EC rather than member governments) would be both better and more politically realistic.

Others responded that the central objective has to be to alter price structures to affect behaviour, not to generate investment programmes - which should be considered separately on their own merits out of central government or other funds. Some argued that the proposed level EC tax was too low to affect behaviour outside of industry, thus returning the discussion to the effectiveness of price incentives.

Proponents noted that revenue-neutrality would still allow the tax burden to be restructured to offset distributional and macroeconomic impacts of a carbon tax. It was suggested that this could be part of a wider reorientation of the tax system away from taxing social 'goods' (such as labour) towards taxing 'bads' (such as pollution). The reduction of other taxes could also be focused to help offset additional costs incurred by industry, without removing the incentive to reduce energy consumption.

Discussion also addressed more general positive industrial aspects of limiting emissions. It was suggested that increased conservation efforts would provide a booming market for insulation materials, and in the longer term, for a range of new industries such as renewable energy technologies. It was also argued that there are considerable long-term industrial advantages in being a 'first mover' in recognizing and reflecting environmental constraints, because it provides a competitive advantage when other countries and regions find they have to adopt similar environmental measures.

Despite disagreements over carbon taxes, most participants agreed that a mix of instruments would be appropriate for limiting CO_2 emissions. Existing experience demonstrates that fiscal incentives have most influence on consumer behaviour in combination with other measures, such as regulation and the dissemination of information. The Commission's SAVE Programme, which was designed to promote energy conservation, but which was 'reduced to the bare minimum of its possible effectiveness', was regarded by many as having a potentially very important role to play. The apparently universal support at the workshop for the SAVE Programme raised the question of why it had been so weakened in the EC. Some attributed this to the increased emphasis upon subsidiarity in EC debates, and/or the dominance of ideology which opposed regulatory measures. Others suggested that it had been weakened precisely because it became confused with the CO_2 issue, and was 'traded off' in an effort to preserve some semblance of agreement on the carbon tax; whereas if it had simply been an energy conservation measure, it might have been approved.

Transport

The transport sector will be the most difficult to deal with in terms of limiting CO_2 emissions. It was widely believed that in the near future there are no real technological alternatives to existing transport structures (although one participant challenged this), and that the area is politically very sensitive.

One suggestion was that the response should be to try to instigate a 'quantum leap' in thinking about transport and environmental problems, not just about CO_2, and that this change should be towards focusing on demand-side aspects of transport. In other words, we should rethink the assumption that demand for transport will necessarily increase, for instance a large share of growth of the demand for transport of persons could be replaced by video-communications. It was accepted that this remains problematic, with little evidence to date that such a shift is occurring or will occur. Indeed in the EC context the Single Market goals increase transport requirements. Motorways have traditionally received a high priority in transport investments, although in some Member States the investment in mass surface transport facilities now equals that in motorways.

It was recognized that the $10/boe tax proposed by the EC Commission might have a very limited effect on the transport sector. One obstacle is that consumers do not adequately weigh the costs of fuel consumption when buying cars. An excise tax on new cars, differentiated by fuel efficiency at the level of ECU 2000 per litre/100km, had therefore been proposed concurrently with the carbon/energy tax. This would have been a significant incentive at the point of purchase towards buying more efficient cars. Considerable support for such a measure was expressed.

Other incentives could be developed. At present, many people have free or subsidized parking at work, paid for by their employers. Altering these incentives could be used to encourage switching to public transport.

However, while much of the above might be desirable, agreement on fiscal measures in the transport sector may be no easier to reach within the EC than in other areas. For example, Germany is very much in favour of road taxes, while Greece is strongly opposed. To get round some of these problems, the Commission has proposed a framework for user charges on motorway heavy transport. These would be designed both to recover the infrastructure costs of motorway building, and to reflect the external (eg. environmental) costs of such transport.

The idea that fiscal measures were needed to stimulate greater fuel efficiency was challenged. It was argued that in the German car fleet from the 1960s to 1980s, there was no correlation between fuel efficiency and price, and that, in the US, efficiency increases had been achieved despite low prices. Price rises might be sufficient to increase efficiency, but they are not necessary, and if necessary the efficiency gains could thus be achieved without the burdens imposed by a tax. Consequently, a measure such as the US's Corporate Average Fuel Economy (CAFE) would be a more appropriate mechanism for increasing efficiency than fuel taxes. Another participant added that, in California, it is legislation on emissions standards which is promoting technological developments that may provide alternatives to the existing transport structure.

This argument - that fuel prices are of minor importance - was disputed as being a spurious use of statistics. With rising incomes there has been a general trend towards bigger cars; but comparative statistics showed a high correlation both between high prices and high fuel efficiency, and between high prices and lower car sizes. It was also argued that there would be great difficulty in negotiating some sort of CAFE-type standards in the EC, since the size of cars varies greatly within Europe. Germany, for example, has very large cars.

Developments in this sector are not motivated by climate concerns, but by problems of congestion and urban air pollution. Dealing with these problems will also limit CO_2 emissions, as for example in Italy where conditions are leading to greatly increased demand for small and diesel cars.

Session 4 - EC implementation: options and constraints

Discussion in the final session centred upon the options for EC implementation and the constraints on EC activity, and covered two areas: how to meet the EC target and what would be the alternatives for a tax regime.

It seems doubtful that the combined plans of all the EC Member States will achieve 1990-2000 stabilization in the EC as a whole if there are no other additional EC actions. It is therefore appropriate to think ahead about the EC's role; but there is still uncertainty about EC competence on global warming policy. Some measures are most appropriately taken at the national level, for example with building regulations. But other measures would have to be implemented at the EC level, notably efficiency standards for traded goods such as cars and appliances. Fiscal measures could be considered at either level; it might be desirable to harmonize these, but it is not essential.

There is a definite risk of finding in December 1993, when all the national plans are to have been submitted to the Commission, that difficult choices have to be made to meet the shortfall. There would then be various options for the Community.

First, the existing unilateral national emission goals could be used as a basis for opening negotiations on explicit 'target-sharing' among member countries. It was suggested that there was strong political resistance to this, and that - even if negotiations were attempted - agreement on the division of binding national emission targets would be very difficult. Secondly, Member States could agree to go back and reassess and strengthen their national strategies. This however may be much the same as trying to negotiate national targets.

The third approach would be to revive and strengthen EC-wide measures, such as the harmonized CO_2 tax and fiscal measures to improve car fuel efficiency. As experience has shown, negotiating effective measures in this approach could also be very difficult.

A fourth option might be for the Community to try to introduce a system of tradeable quotas in some form. The national flexibility afforded by allowing tradeability might make negotiations easier than attempting to negotiate fixed national targets, or a substantive package of EC-wide measures. The advantage is that it would be a consistent EC-wide approach, but in accordance with subsidiarity, competence to enact specific abatement policies within this system would remain with member governments. It would however raise problems, not least the novelty of such an approach in the EC.

This raised wider questions as to what a fair allocation of responsibility might be. Since it had been asserted that negotiating binding emission targets for each Member State was not a realistic option, and that whilst tradeability might ease some of the difficulties it would still be extremely difficult, it was suggested that an alternative approach might be to establish criteria for a fair allocation of efforts towards limiting emissions, rather than of emissions themselves. Some agreed that this might be fruitful, but others considered that in practice this was little different from the current proposed process of national strategies and their review by the Commission.

Alternative outcomes were suggested. One is the possibility that the Commission simply shrugs its shoulders and says 'dommage'. This was not expressed as a desirable outcome, but would not be too problematic legally, since the Convention commitments are phrased so as not to be legally binding. Alternatively the EC might find ways of amending the statistical basis of stabilisation (for example, concerning certain statistical definitions or geographical scope relating to EC expansion during the period). Perhaps the most likely outcome is that the Commission tries to make up the shortfall in emissions reductions by a combination of further harmonized EC-wide measures, and by looking at the existing plans, identifying possible areas for improvement, and negotiating bilaterally with Member States to alter plans in line with these potential improvements.

Discussion finally returned to aspects of the carbon tax related to national differences and cross-border implications. An objection was raised that it would have very differential impacts across the EC, because the fuel mix differs widely between Member States. Conditions also differ in other ways; for example, in Germany, the tax may have little effect on fuel switching because substantial investments have recently been made in clean coal technology. Furthermore it was suggested that a tax would not have the benefit of EC-wide efficiency claimed for it because there is not a free market for energy in Europe. Many European energy industries, particularly utilities, are vertically integrated, monopolistic and protected. It was suggested that this would limit the effect of a tax on fuel switching, a limitation which would be compounded by existing subsidies, such as the Kohlenpfennig ('coal penny') in Germany.

It was recognized that Member States would be free to decide the most appropriate ways of recycling revenues from a carbon/energy tax in their country, ostensibly within the constraint set that the tax would be revenue neutral, ie. offset against other tax reductions so as not to increase the aggregate tax burden. There was some scepticism expressed, not only about the desirability of such fiscal neutrality (as noted above), but about the feasibility of monitoring it. Many participants considered that in practice such taxes would be introduced primarily with a view to raising revenue and reducing budget deficits.

A final political problem with the proposed tax was raised from the perspective of the EC. Community institutions would be seen as the villain for introducing a new tax, whilst Member States would get the revenue and any political benefit from reducing other taxes.

Part II: Papers to the Workshop

EC climate policy and the UN Framework Convention on Climate Change

Michael J. Grubb
Senior Research Fellow
Royal Institute of International Affairs
London

Abstract

The European Community has sought to lead international efforts to address climate change. The processes leading to the UN Framework Convention on Climate Change, and the agreement itself, have strongly reinforced the EC commitment to stabilize total CO_2 emissions. It also establishes an international process which requires both individual members, and the EC itself, to report the details of their strategies for achieving emission goals, including reappraisal of progress and of policies which impede emission reductions.

Several of the less developed EC member countries agreed to stabilization on the understanding that increases in their emissions will be allowed and offset by reductions in other EC countries, and they entered into the Climate Change Convention on the same basis, under the provision for joint implementation. For higher-emitting northern EC countries, a goal of national 1990-2000 stabilization thus appears secondary to the reductions implied by the EC commitment, and the EC signature to the Convention. If both member governments and the EC are to report to the Convention's Conference of Parties as required, 'burden sharing' arrangement will have to be clarified if these reports are to be consistent, and clear decisions will have to be reached on the application of subsidiarity and policies such as coal subsidies. The EC probably has less than two years in which to achieve and implement such agreement.

Introduction

Since concerns about human-induced climatic change first emerged as a major political issue in the mid to late 1980s, EC governments have been in the forefront of efforts to start addressing the problem. Some EC countries, and later the EC itself, adopted unilateral emission goals and sought to follow these through with policies to limit particularly CO_2 emissions. Building upon this, the EC sought to lead the international process established by the UN General Assembly towards a strong Framework Convention on Climate Change.

The Convention which emerged from this process, signed by 153 countries at the UN Conference on Environment and Development at Rio in June 1992 (and by several more since), now forms the legal basis for the international development of responses to the climate problem. In this presentation I discuss issues raised by efforts to implement the EC's commitments and how they relate to, and are affected by, the Climate Change Convention.

Three areas of the negotiations and resulting Convention are of particular relevance to this: the logical basis for precautionary action to limit emissions; the general nature of North-South responsibilities; and the specific nature of commitments and policy responses under the convention.

Precautionary action

When the EC Council of Ministers first agreed the EC's goal of stabilizing CO_2 emissions, it was a unilateral initiative which reflected the widespread view that the dangers of climate change required significant action to be taken, notwithstanding the acknowledged uncertainties surrounding the issue. Two specific aims were to inject into the climate negotiations both the principle of precautionary action, and the objective of CO_2 emissions stabilization by industrialized countries as an appropriate initial response.

On both counts, with minor caveats, the EC succeeded. The Objective of the Convention is agreed to be:

> ... stabilization of greenhouse gas concentrations in the atmosphere at a level that would prevent dangerous anthropogenic interference with the climate system ... within a time frame sufficient to allow ecosystems to adapt naturally to climate change, to ensure that food production is not threatened and to enable economic development to proceed in a sustainable manner.

The Principles then state that:

> The Parties should take precautionary measures to anticipate, prevent or minimize the causes of climate change and mitigate its adverse effects. Where there are threats of serious or irreversible damage, lack of full scientific certainty should not be used as a reason for postponing such measures,

taking into account that policies and measures to deal with climate change should be cost-effective so as to ensure global benefits at the lowest possible cost.

Both these statements are based upon wording as formulated by the EC. The acceptance of the 'precautionary principle' (albeit in this mild form) reflects the fact that attempts to argue that all action should wait upon greater scientific certainty proved logically unsustainable.

The qualifier of 'cost effectiveness' indicates that countries did not regard addressing climate change as an over-riding priority, to be addressed with an unlimited purse. Its meaning is of course extremely ambiguous, and it gives wide scope for future disputes over its interpretation. Nevertheless, the quantified aim for industrialized countries to return emissions of CO_2 and other greenhouse gases to 1990 levels by 2000 was largely accepted as a suitable goal, as discussed below.

North-South roles and responsibilities

Developing countries varied greatly in their detailed attitudes in approaching the Climate Convention, but all were united in the underlying perception that it was a problem for which the developed countries bore primary responsibility, because of their greater wealth and technical capability and because of their much greater past and current emissions. Developing countries made it plain that they could not consider taking action until developed countries had started to take significant measures, and that they could not be expected to incur additional costs in abatement without compensation.

Most of the developed countries accepted the logic of this position, and many European countries especially were sympathetic to the concerns and position of developing countries. Accordingly, concerning *principles*, the Climate Convention acknowledges that responses should be:

'on the basis of equity and in accordance with [States'] common but differentiated responsibilities and respective capabilities. Accordingly, the developed country Parties should take the lead ... '
[with particular recognition of] 'the specific needs and special circumstances of developing country Parties, especially those that are particularly vulnerable .. '

Correspondingly, the commitments for developing countries are looser, and developed countries

'shall provide new and additional financial resources to meet the agreed full costs incurred by developing country Parties' in preparing their reports under the convention, and 'shall also provide such financial resources, including for the transfer of technology, needed by the developing country Parties to meet the agreed full incremental costs of implementing measures that are covered by paragraph 1 of this Article and that are agreed [between that country and the funding agency].'

The Global Environmental Facility (GEF), pioneered by France and Germany, is established in the Convention as the interim funding agency. The net effect of these outcomes concerning the North-South responsibilities is to place additional emphasis upon the need for developed country action as a precursor to getting developing countries seriously involved, and also to place greater weight on the EC/EFTA stance concerning this and the need for, and mechanisms for, assisting developing country participation through the GEF. The detailed implementation of these developments are not examined further in this paper, which concentrates upon internal EC policy.

The stabilization goal

Negotiations on the Climate Convention amongst OECD countries were dominated by the dispute between the US and most others, centred upon the EC position, about the adoption of emission targets. These latter countries, having adopted unilateral emission goals (mostly relating to stabilisation or tougher targets for year 2000 CO_2 emissions), sought to include such a commitment for all developed countries. The US argued that such a commitment was arbitrary, unenforceable and premature.

In the end, the UK brokered a compromise which clearly retained the objective of 1990-2000 stabilization as a guideline of desirable action, without being a legal commitment. Specifically, two disjointed references appeared, one to the desirability of returning emissions 'by the end of the present decade to earlier levels', the other a specific 'aim of returning [emissions] .. to their 1990 levels' (for detailed wording see box).

The view taken by most EC countries - and one which now seems bound to be accepted by the new US Administration - is that despite the deliberate ambiguities, this is to be interpreted as a commitment by all the industrialized country Parties to return CO_2 emissions to 1990 levels by the year 2000, individually or jointly. It is also clearly the view taken by developing countries, who make it plain that they have no intention of acting whilst the much higher emissions from developed countries are still rising.

It is important to note the status which the whole process and its outcome has lent to the EC stabilization commitment. To renege upon it now would undermine the UN Convention in the very area in which the EC fought hardest for stronger wording, and would be used by developing countries as a *prima facie* reason why they should not take significant action. Given that the target has already been reaffirmed by the EC Council of Ministers, it would also make the EC look foolish, if not actually devious, and lacking in seriousness about

Commitments for developed country Parties to the UN Framework Convention on Climate Change.

a) 'Each of these Parties shall adopt ... policies and take corresponding measures on the mitigation of climate change, by limiting its anthropogenic emissions of greenhouse gases and protecting and enhancing its greenhouse gas sinks and reservoirs. These policies and measures will demonstrate that developed countries are taking the lead in modifying longer-term trends in anthropogenic emissions consistent with the objective of the Convention, recognizing that the return by the end of the present decade to earlier levels of anthropogenic emissions of carbon dioxide and other greenhouse gases not controlled by the Montreal Protocol would contribute to such modification ...'.

This is qualified by '... taking into account' the need to recognise various differing circumstances, and to 'maintain strong and sustainable economic growth', and states that 'Parties may implement such policies and measures jointly with other Parties'.

b) 'Each of these Parties shall communicate, within six months of the entry into force of the Convention for it and periodically thereafter ... detailed information on its policies and measures referred to in subparagraph (a) above, as well as on its resulting projected anthropogenic emissions .. for the period referred to in subparagraph (a) [ie. to the year 2000], with the aim of returning individually or jointly to their 1990 levels these anthropogenic emissions .. This information will be reviewed by the Conference of the Parties, at its first session and periodically thereafter, in accordance with Article 7';

c) '.. The Conference of the Parties shall consider and agree on methodologies for these calculations at its first session.'

d) 'The Conference of the Parties shall, at its first session, review the adequacy of subparagraphs (a) and (b) above ... [and] take appropriate action, which may include the adoption of amendments to the commitments ... A second review ... shall take place not later than 31 December 1998 ...'

This section further requires each developed country Party to:

'coordinate as appropriate with other such Parties, relevant economic and administrative instruments...', and

'identify and periodically review ... policies and practices ... that lead to greater levels of emissions ... than would otherwise occur'

such undertakings. As noted in another paper in this workshop (Haigh), the proposed Council Decision currently before the Council[18] would have the effect of turning this political commitment into a fact of Community law. Accordingly, there are now very strong pressures to ensure that the EC stabilization goal is not abandoned, and to find ways of achieving it.

There is however an extremely important aspect of this which has not been adequately recognized, at least in the UK. The Convention is signed by all EC member countries, *and* by the EC as a separate entity. But stabilization by the EC is not the same thing as stabilization by all individual states, since the former allows emissions by some countries to increase *if* others reduce accordingly. Specifically, in agreeing to the EC stabilization goal, several countries on the EC 'periphery' - notably Spain, Portugal, Greece and Ireland, which start from a base of per-capita emissions far below the EC average - made it plain that they would *not* stabilize their own emissions. They participated under assurances that increases in their emissions would be allowed and offset by reductions in emissions from some of the more developed members of the Community. They signed the Convention on the same basis, drawing on the provision for joint implementation of the stabilization goal - namely, their participation in the EC goal - as their commitment.

The higher-emitting northern EC countries have thus entered two commitments. One is directly to the Convention. The other - direct to the EC and indirectly to the international community through the EC participation as a signatory of the Convention and the provision for joint implementation - is joint EC stabilization. For the northern EC countries, this may make national stabilization redundant, because achieving it will require them actually to reduce their emissions.

Policies, reporting and review procedures

One of the underlying tensions between the US and the EC positions in the negotiations was that the US doubted the EC's ability to deliver emission stabilization, and was not convinced that the policy requirements and implications of achieving such goals were adequately understood. The US persistently took a stance that was directed more towards the need to explore and implement policies, and less towards emission goals. The wording on commitments displayed in the box above, whilst reflecting the stabilization guidelines, is very much couched in these terms: the emphasis is upon adopting policies and reporting on them, and their impact on emissions, to the Conference of Parties.

Furthermore, and partly as the price for not obtaining a binding commitment to emissions stabilization in the Convention, the EC was keen to ensure a strong mechanism for reviewing the adequacy of commitments in the convention, whilst the US was also keen to see that policy pronouncements were examined

---

18. See the Postscript to this report.

seriously. Consequently, the Convention contains relatively strong procedures to ensure both national reporting on measures adopted, and review of the adequacy of commitments.

As a result of this, all developed country Parties - including both EC member countries *and* the EC as an entity - are required to submit to the Conference of Parties national reports on the policies adopted towards stabilization, with projections of emissions. Again, this has clear implications for EC climate policy. If the EC is to meet its commitments under the Convention, it must submit a report detailing the policies adopted, with projections to show that these will achieve the stabilization goal, and national reports and projections which should be consistent with this. In principle, therefore, all the major issues relating to the stabilization goal, including the EC policy instruments used, the arrangements for burden-sharing, and associated national projections (or an equivalent EC-wide market-based system) must be agreed by the time of submission of the report.

The Convention will enter into force after the fiftieth ratification. At the time of writing it appears likely that this will be achieved by mid 1994 at the latest. Consequently, it is likely that a comprehensive EC report, detailing strategy in place to achieve stabilization, must be lodged by late 1994.

The EC policy dilemma

This is significant because of the current major uncertainties surrounding EC climate policy. Having decided very early on that an arrangement of explicit burden-sharing through national emission targets was not feasible (and not appropriate to a converging Community), the member governments asked the European Commission to prepare a Community-wide strategy to stabilize EC CO_2 emissions. Of the resulting 5-part strategy:[19]

- the energy/carbon tax is recognized to be in great difficulty and clearly will not be implemented on the proposed timescale

- the SAVE programme of energy efficiency standards has been largely sacrificed on the altar of subsidiarity, with an understanding that member states are free to pick and choose measures, subject to EC competition law

- the THERMIE programme for promoting energy efficient technologies through demonstration and enhanced diffusion schemes can only make a marginal contribution

---

19. 'A community strategy to limit carbon dioxide emissions and to improve energy efficiency', COM(92) 246 final, Brussels 1 June 1992.

- the ALTENER programme for promoting renewable energy technologies likewise can have little impact by 2000, because of inherent timing constraints and lack of funding

- the result is to place nearly all the weight upon the fifth component of the strategy, namely the 'monitoring mechanism' proposals by which member countries develop and submit to the Commission their own national strategies for abatement - which presumably have to subsume the original SAVE proposals and considerably more.[20]

Member countries are currently required to submit to the Commission reports on their national programmes by the end of 1993. What will happen if the resulting national projections do not add up to stabilization - as seems very likely - is then wholly unspecified, save that then 'the Commission may, if necessary, and with due regard to burden sharing, make all appropriate proposals for required additional actions .... '.[21]

What form might 'additional actions' take, when the actions proposed by the Commission have been so greatly weakened? I have argued elsewhere that the most desirable and convincing approach would be for the EC to return to a measure of target-sharing, but with the crucial difference that the targets would be tradeable between member states.[22] It is possible that the EC anyway will be driven down this road by the force of positions already adopted, despite the fact that most member countries are nominally hostile to the idea of such a 'tradeable targets' approach.

What is certain is that neither this, nor any other major new development in EC strategy which may be required if emissions stabilization is to be ensured, can come about easily or quickly unless there is much forethought and discussion which recognizes that the current strategy is probably inadequate. If the EC waits until the end of 1993 in the hope that national strategies have been developed that can achieve the goal, this will leave barely a year until the EC has to submit a report detailing its policies and emission projections to the UN Conference of Parties. If the EC as a body is to meet the commitments it has undertaken, it may thus have less than a year to propose new and additional measures and to have them accepted throughout the EC.

This scarcely looks feasible: partly because of the intrinsic difficulty of getting policy agreed in such timescales; partly because of the inevitable political difficulties raised by burden-sharing; and partly because more funda-

---

20. Proposal for a Council Decision for a monitoring mechanism of Community CO2 and other greenhouse gas emissions, SEC(92) 854 final, Brussels, May 92.

21. ibid.

22. M.Grubb and C.Hope, 'EC Climate Policy: Where there's a will ....', *Energy Policy*, Butterworths, November 1992.

mental questions have yet to be settled. Two which feature highly are the application of subsidiarity concerning, for example, efficiency standards on tradeable goods, and the EC attitude on policies such as the German coal subsidies and electricity trade. Unless there is more rapid forward thinking and preparation, EC climate policy thus appears to be heading for considerable trouble, which will be both internationally embarrassing and potentially damaging to the processes established by the Climate Convention.

Implementation of the German CO_2 reduction goals

Dr Edda Muller
Federal Ministry for the Environment, Nature Conservation and Nuclear Safety
Germany

I should like to take the opportunity here of reporting on the state of implementation of the Federal government's CO_2 reduction concept.

1. Point of departure

You are aware, I am sure, that the Federal government has now taken three decisions pertaining to reductions in CO_2 emissions in the Federal Republic of Germany. These decisions contained not only an ambitious objective, but took the form of a far-reaching catalogue of measures now being gradually implemented.

This catalogue of measures contains both economic instruments and instruments under administrative law - so-called command and control measures - backed up by 'soft' measures such as information, consultation, education and training. Parallels may be drawn at all levels between this package of measures and the Community's strategy to reduce CO_2 emissions and improve energy efficiency submitted by the EC Commission on 25 October 1991.

But let's come back to Germany's CO_2 reduction policy.

It remains the Federal governments's objective to reduce energy-related CO_2 emissions by 25-30% by the year 2005 taking 1987 as the base year. This is without doubt a very ambitious objective, particularly given the forecasted rise in real gross national product by around 50% - an average annual rise of 2.5%. In other words, it is our intention to produce one and a half times the goods and services provided today with just 70-75% of 1987 CO_2 emissions.

2. The state of the implementation of the Federal government's CO_2 reduction programme

The following measures which help reduce CO_2 are already in force:

(1) The amended Federal Code for Electricity Price Rates, giving financial incentives for rational and economical electricity use in private households, in trade and industry and in agriculture.

(2) 250MW wind energy programme promoting demonstration facilities.

(3) The 1000-roof photovoltaic programme promoting small, decentralized photovoltaic facilities operated privately.

(4) The Act on the National Grid by which electricity companies are obliged to take electricity from independent renewable sources into the national grid at a price geared towards average electricity prices (wind/photovoltaic 16.53 pfennigs per kilowatt-hour).

(5) A programme undertaken jointly by the Federal government and the Länder over a period of several years to modernize and clean up district heating networks in the new Länder, with particular emphasis on cogeneration. A yearly backing of 300 million deutschmarks is set to mobilize an annual investment of around 1 billion marks.

(6) Tax incentives for cogeneration prescribed in the Mineral Oil Act.

(7) Programme to implement energy diagnoses in the construction sector with a 900 marks grant on every energy diagnosis carried out.

(8) Support for consultation for small and medium-sized companies to optimize their energy provision.

(9) ERP credit programme to provide low-interest loans to promote rational and economical energy use and the application of renewable energies in small and medium-sized companies.

(10) The environmental label used to provide customer information on rational and economical use of energy and the application of renewable energies, for example for

— low-emission and energy-saving refrigerators
— low-emission oil atomizer burners
— gas and special heating boilers
— combined water heaters and circulation water heaters for gaseous fuel
— burner-boiler units with gas burners fitted with a ventilator
— oil burner-boiler units
— solar-powered products
— low emission and energy saving gas boilers
— gas room heaters and gas heating facilities.

Now to the measures to be implemented over the next few months:

(1) The amended Heat Protection Ordinance demanding a low-energy standard (50-90kWh/m^2/annum) for new buildings and the gradual reaching of this standard in existing buildings.

(2) The amended Heating Facility Ordinance (applying to heating facilities of between 4 and 400kW) designed to ensure the now advanced state-of-the-art technology.

(3) The amended Ordinance on Small Firing Facilities (applying to small firing facilities of between 4kW and 1MW for coal, 5MW for oil and 10MW for gas) to make the provisions more stringent in this area too (eg. introduction of gross calorific value technology for new facilities).

(4) Submission of a Heat Use Ordinance obliging operators for larger firing plants either to use the waste heat generated in their plants in actual operation or to supply it to third parties for use in their plants or in the district heating systems.

(5) The amended Energy Industry Act dating from 1935 and regulating mains-based fuels. The objective of the amendment is to take greater account of aspects of environmental protection and the preservation of resources and to implement provisions decided on at European level.

(6) The amended Pay Code for Architects and Engineers giving economic incentives to architects to plan with energy efficiency in mind and to ensure corresponding implementation.

Moreover, the Federal government believes that the raising of a CO_2/energy tax is absolutely essential in order to reach its set objective. From the point of view of the global problem of the greenhouse effect, too, Germany supports the proposal of the EC Commission to introduce an EC-wide combined CO_2 and energy tax. However, in harmony with other government bodies, the Federal Ministry for the Environment, Nature Conservation and Nuclear Safety rejects any conditions put on this tax (it should not depend on the introduction of comparable measures in the countries of the OECD).

With regard to the use of economic instruments, the Federal government is at the moment joining with trade and industry to look into how compensation - joint commitments - might work to structure CO_2 reduction measures as cost-effectively as possible (bubble concept).

Over and above these measures, the Federal government has made a whole range of other provisions contained in the text of the cabinet decisions. More-over, the inter-ministerial working group on CO_2 reduction set up by the Federal cabinet on 13 June 1990 under the responsibility of the Federal Ministry for the

Environment, Nature Conservation and Nuclear Safety is continuing to fulfil its remit, with consultations on the structure of the CO_2 reduction programme. This working group is due to submit a further report on CO_2 reduction to the cabinet in autumn 1993 and, if need be, propose new measures.

3. The role of the Länder, the local authorities and trade and industry
To achieve the set CO_2 reduction objective, it is obvious that in a federal system, the reactions of other levels of administration are of immense importance. Concerning this:

(1) The Länder of North Rhine-Westphalia and Schleswig-Holstein have already submitted CO_2 reduction concepts specific to their particular area. Other Länder such as Brandenburg, Saxony, Lower Saxony, Baden-Württemberg and the Saarland are also preparing programmes of this kind.

(2) More than 50 local authorities listed in the attached paper have also taken the opportunity presented by the decisions of the Federal government to look into possible ways of reducing CO_2 on the basis of extended local and regional energy supply plans and define measures to realize them.

German trade and industry too has expressly stated its readiness to play a part in reducing CO_2. In November 1991, the central federations of German trade and industry submitted an initiative for preventive action on the climate, designed to produce intensive discussions with representatives from all areas of the economy, to sound out possible self-commitments and consider future cooperation.

Already, the Federal government's reduction programme is proving, in certain sectors of the economy, to be an incentive for a deep-seated restructuring and modernization of production potential, thereby leading to improved competitiveness for the German economy in the medium to long-term. And finally, suppliers of energy efficient techniques, products and services are also involved - given growing demand - in helping solve the problems entailed by CO_2 reduction.

4. The development of CO_2 emissions in Germany between 1987 and 1991
Data published by the Federal Ministry for Economics shows a more than 11% drop in CO_2 emissions from the united German territories between 1987 and 1991. Calculations made by the Federal Ministry for the Environment, Nature Conservation and Nuclear Safety estimate the reduction in CO_2 in Germany between 1987 and 1992 at around 14%. The drop is primarily due to the collapse of the economic structures in the new Federal Länder. To consider the small reduction in energy consumption in the western Länder as proof that developments are running counter to the CO_2 reduction objective would, how-

35

ever, be completely wrong since the vast majority of goods and services required in the new Länder are produced in the old.

The development in primary energy consumption in the first half of 1992 also underlines our belief that the twenty-year-old trend towards restructuring and optimization of production and consumption is continuing and the trend of CO_2 emissions is still downwards.

5. Summary

It would, however, be too optimistic to conclude from this information that the German CO_2 objective can be reached without problems. Take the traffic sector as an example.

Traffic is at the moment our greatest concern. Following the collapse of the communist systems of the east, the Federal Republic of Germany is developing more and more into the pivot of European transport, with innumerable negative consequences both for transport policy and the environment. This is true particularly for traffic-induced CO_2 emissions, and of course also for increased outputs of NO_x, VOC, heavy metals, particulates and soot. An overall concept must be developed here in cooperation with our European neighbours in east and west, north and south, to prevent a situation arising over the next few years requiring massive intervention in traffic and environment policy.

UK policy toward climate change

Alan Davis
Department of the Environment
United Kingdom

The commitment to devise, implement, publish and regularly update national plans or programmes may well turn out to be the most significant part of the Convention; coupled with reporting and review machinery it represents a major advance in global action to tackle climate change.

The UK has already published a brief description of its strategy in response to a call from the Environment Council and also announced in the Second Anniversary White Paper report that we will be issuing a wide ranging discussion paper which will set out in detail the options for further measures to limit CO_2 emissions.[23] This is aimed at industry, commerce and the public, to build a consensus for the kind of steps we will need to take to meet our Convention commitment.

The G7 Munich Summit in July 1992 committed major industrialized countries to seek to ratify the Convention by the end of 1993; the USA and three island countries had ratified the Convention by mid-October 1992. The UK is keen to see it come into force as soon as possible and this means we must get on with ratification by all EC countries.

Before Community and Member States can ratify, we will need to decide how the Convention commitment on CO_2 emissions relates to the conclusions of the October 1990 Environment/Energy Council. These set a conditional aim for Member States of taking actions aimed at reaching stabilization of the total CO_2 emissions in the Community as a whole at 1990 levels by 2000.

In practice, measures to meet the Convention legal commitment will be the same for the next few years as those we would take towards meeting the October 1990 aim; the key is for all Member States to produce national plans describing measures they are taking and their impact on CO_2 emissions.

It is likely that this will show that the Twelve are meeting the Convention commitment of action aimed at returning CO_2 emissions to 1990 levels in 2000; in that case, we can agree to meet this part of the Convention commitment jointly.

National plans for CO_2 emissions are of course only one part of the commitment to plans and programmes under the Convention. These will need to cover actions on other sinks and sources, eg. on forests as sinks for CO_2, and on all other commitments in the Convention. It is important, if we are to meet the

---

23. This was published in December 1992 as *Climate Change: A Discussion Document*, HMSO, London, 1992.

commitment of publishing full plans by the end of 1993, that we each focus on producing comprehensive and effective plans.

This will demonstrate to developing countries that we are taking the Convention seriously and mean to implement commitments; we need to encourage them to come up with detailed and effective plans which enable the GEF as the Convention financial mechanism to fund projects to limit emissions and protect sinks.

It will be important that we work closely together to develop national plans. The OECD may well have a role to play as a forum for bringing together experts to compare notes on measures we are all looking at. It will be useful to draw on different experience to help all of us produce the best and most effective balance of measures.

Strategies and perspectives towards climate change in Spain

Luis Carlo Mas Garcia
Direccion General de Politica Ambiental Ministerio de Obras Publicas y
Transportes
Secretaria de Estado para las Politicas del Agua y el Medio Ambiente
Spain

The Spanish strategy in response to the problem of climate change has been
based on two main points of reference over this past year. Firstly, with a
medium and long-range perspective, the National Climate Commission has been
created. Secondly, in answer to the needs of the Community strategy, the
National Programme for the Limitation of CO_2 Emissions has been drafted. In
following, a brief review will be provided of these two aspects of the Spanish
strategy.

National Climate Commission
By way of the Royal Decree of May 29, 1992, the Spanish government has
established the National Climate Commission (Comision Nacional del Clima).
The main objective of this body is that of acting as a government policy adviser
regarding climate change and the pertinent response strategies.

The Commission has been organized under the Ministry of Public Works and
Transports, through the Secretary of State for Water and Environmental Policies.
The functions set for the National Climate Commission are as follows:

- to collaborate in drafting and developing a National Climate Programme
- to harmonize the activities of that Programme with other National Plans and
Programmes that may have to do with climate change
- to promote informative, publication and training activities related to the
climate and climate change
- to act as an adviser for the Spanish delegations to intergovernmental
agencies
- to analyze and recommend the measures to be taken by the government as
required for the purpose of fulfilling the commitments undertaken in international conventions and protocols and the pertinent measures to be taken in
the sectors affected by climate change
- to coordinate the drafting of national climate change reports.

Under the Chairmanship of the Minister of Public Works and Transports, the
following Departments are also represented on this Commission: the State
Department; the Treasury Department; the Department of the Interior; the
Education and Science Department; the Department of Industry, Trade and

Tourism; the Department of Agriculture, Fishing and Food and the Office of the President.

National programme for the limitation of CO_2 emissions
Prior to the setting up of the National Climate Commission, the Spanish government began to work on drafting a National Programme for the Limitation of CO_2 Emissions. This was pursued especially following the meeting of the Community Council of Ministers of Energy/Environment held on December 13, 1991, the conclusions of which proclaimed the need for the Member States to draft such programmes and forward them to the European Commission.

The Spanish Programme has been coordinated jointly by the Environmental Policy Directorate of the Ministry of Public Works and Transports and the General Secretariat of Energy and Mineral Resources of the Ministry of Industry, Trade and Tourism and has received the input of other Ministerial Departments. The Programme was forwarded to the Commission in July 1992.

The Spanish National Programme for the limitation of CO_2 emissions is based on the National Energy Plan (NEP) 1991-2000 which received government approval in 1991. The main instrument that the NEP sets forth for rationalizing and limiting energy consumption and, thus, CO_2 emissions, is the Energy Saving and Efficiency Plan (ESEP).

The ESEP includes a number of measures having an influence on both final energy demand (less power used whilst maintaining the same level of economic activity) and on promoting and putting new power variations into use (cogeneration and renewable energy sources).

The end purpose of the Energy Saving and Efficiency Plan is to reduce the expected level of final energy demand by 7.6% of the projected level by the year 2000, which means decreasing the demand as estimated without any measures being taken by 6.3Mtoe.

The ESEP is divided into four programmes of measures: Saving, Substitution, Cogeneration and Renewable Energy Sources.

Saving Programme
The goal of this Programme is that of limiting final energy consumption without affecting the level of the economic activity. The main measures planned, by sectors, are as follows:

Industry: In 1990, this sector represented 40.3% of all final energy consumption, and it is estimated that its share therein will be reduced to 38.5% by the year 2000. The measures to be used involve technical aspects (using new technologies), management (improvement of monitoring and training methods) and consumer information and education.

Transport: In 1990, this sector represented 37.4% of all final energy consumption and 55% of the final petroleum-product consumption, it being estimated that this will be reduced to 39.2% of the final consumption by the year 2000. The measures planned are equivalent to those for the industrial sector.

Other uses: In 1990, other sectors represented 22.3% of all final power consumption, and it is foreseeable that it may remain almost the same for the year 2000. In this case, measures will be taken regarding residential buildings, utilities, administration and street lighting.

The impact of this Energy Saving Programme on final demand and CO_2 emissions, is shown in Table 1.

Table 1. Projected energy and CO_2 savings by 2000 from Spanish Energy Saving Programme

Energy source	Savings by sector, Ktoe*			Total Savings	
	Industry	Transport	Other uses	Ktoe*	Kton CO_2
Coal	-445	--	-10	-455	-2228
Petroleum products	-656	-3135	-511	-4302	-13240
Natural gas	-511	--	-71	-582	-1362
Total				-5339	-16830

*1 Kilotonne of oil equivalent = 7,350boe = 44,600GJ (GHV)

Substitution Programme

The objective of this Programme is to make changes in the structure of energy demand, changing over from oil products and coal to natural gas, thus achieving a reduction in CO_2 emissions, whilst obtaining added savings from reduced capital requirements and greater efficiency, due to improved combustion.

The main sectors and measures planned are as follows:

Industry: Partial substitution of fuel-oil and coal. The industrial sectors in which this changeover is most likely to be successfully made are the cement manufacturing and iron and steel-manufacturing sectors.

Other uses: In this area, expanding the natural gas network is the objective, thus making it possible to replace gas-oils and fuel-oils, home heating being the prime example.

The impact of this Programme on final demand and CO_2 emissions is shown in Table 2.

Table 2. Projected energy and CO_2 savings by 2000 from Spanish Substitution Programme

Fuel	Energy substitution, Ktoe			Total CO_2, $KtCO_2$
	Industry	Other	Total	
Fuel oil	-974	-102	-1076	-3572
Gas-oil	-34	-96	-130	-403
Coal	-113	-284	-397	-1605
PLG	--	-50	-50	-140
Comb. total	-1121	-532	-1353	-5720
Tech. savings	+66	+57	+123	
Total	-1055	-475	-1530	
Natural gas	+1205	+475	+1680	+3930
Total substitution			+150	-1790

Cogeneration Programme

This Programme is aimed at promoting the use of combined production of heat and power. The main measures planned are as follows:

Industry: Focusing on energy intensive industries, especially the refinery, chemical and food (sugar-manufacturing) sectors.

Other uses: The main subsector being focused upon in this case is the hospital sector.

The impact of this Programme is shown in Table 3.

Table 3. Projected energy and CO_2 savings through cogeneration

Energy source	Energy Savings, Ktoe			Total CO_2 savings, $KtCO_2$
	Industry	Other	Total	
Coal	-58	--	-58	-229
Petroleum	-402	-95	-497	-1629
Natural gas	+419	+94	+513	+1200
Total cogeneration			-42	-658

Renewable Energy Programme

This Programme is aimed at promoting the use of renewable and other new energy sources. The energy sources dealt with in the Programme are:

Mini-hydroelectric: Hydroelectric plants of up to 5MW, with the aim of displacing 213 Ktoe of primary energy.

Biomass: The main contribution stems from the direct burning of farming and forestry by-products for industrial use.

Solid urban waste: The incineration of this waste is envisaged with the recovery of the energy produced.

Wind-energy: In the year 2000, the installed electric power will be 150-200MW.

Solar heating and photovoltaics: Both show significant prospects for development in the medium and long term, although their influence is currently quite limited.

Geothermal: Some projects in the area of low-temperature studies are going to be undertaken.

The Renewable Energy Programme as a whole will displace 500Ktoe of final consumption, although given the problems involved in precisely determining the

43

biomass or solid urban waste emissions, their effect on reducing the amount of CO_2 emissions has not been taken into account.

The figures taken into consideration up to this point in the description of the ESEP have estimated solely the savings or the substitution involving coal, petroleum products and natural gas. In addition, electrical power savings must be evaluated with regard to the following two aspects:

- The Saving Programme, which estimates a power saving by the year 2000 of 820Ktoe. In terms of CO_2, a reduction on the order of 9033Kton had been estimated, which would be the amount given off into the atmosphere on generating 820Ktoe of electricity at a combustion plant using imported coal.

- The selection of electrical systems. Added hydroelectric power, importing electricity from France, the use of clean technologies for burning coal and, fundamentally, the option for natural gas, lead to a projected reduction for the year 2000, of 14543Kton of CO_2 emissions below the reference projection.

For the purpose of adequately evaluating this last piece of data provided, it must be stressed that the CO_2 emissions from power generation, without taking into account the National Programme for the Limitation of CO_2 Emissions, would have increased from 1990 to the year 2000 to a much greater extent than the estimated increase in power consumption over the same period of time. This is because the power system structure in 1990 was such that a low degree of carbon dioxide was given off into the atmosphere, due to the large amount of hydroelectric and nuclear power; and precisely this last energy source is subject to a moratorium declared by the government which will mean that its role will be noticeably reduced by the year 2000.

Thus, for the purpose of comparison with the situation were no measures to be taken, it has been considered that new generation would have been from plants using imported coal.

The overall effect of the measures planned in the Spanish National Programme for the Limitation of CO_2 Emissions would mean that Spanish emissions increase would be limited to 25%, from the 218 million tons of 1990 to 272 million tons in the year 2000, instead of the foreseeable 45% increase were the current trend to continue.

Table 4 shows emissions broken down by sectors.

Table 4. Spanish CO_2 emissions by sector, 1990 and projections

	Year 1990		Year 2000 without NEP			Year 2000 with NEP		
	MtCO$_2$	Proportion	MtCO$_2$	Proportion	2000/1990	MtCO$_2$	Proportion	2000/1990
Industry	45.6	20.9%	55.6	17.6%	21.8%	48.5	17.8%	6.4%
Transport	67.1	30.8%	96.3	30.5%	43.6%	86.7	31.8%	29.3%
Others[1]	25.4	11.6%	29.9	9.5%	18%	27.3	10%	7.5%
Processing[2]	79.9	36.7%	133.6	43.4%	67.26%	110	40.4%	37.7%
Total	218	100%	315.5	100%		272.6	100%	
Index	100		145			125		

[1] Primary, domestic and tertiary sectors
[2] Power, refinery and coal

Currently, Spanish emissions in per capita terms are approximately 65% of the Community average, the lowest relative level other than Portugal. Following the 25% increase planned in the Programme and supposing that an overall stabilization is achieved at the Community level, in relative terms, per-capita emissions in Spain would be 82% of the Community average. Under these conditions, the objective that the programme sets forth for Spanish CO_2 emissions can be deemed reasonable.

It is the hope of the Spanish administration that, from this point on, the European Commission will be continuing the logical process initiated when Member States began drafting their National Programmes. That is to say, that once the programmes of all the nations have been presented, the Commission will have to evaluate the overall effects thereof and what they will mean in terms of the overall carbon dioxide emissions in the Community. Were this consolidation of the National Programmes to suffice to achieve the objective of Community stabilization, the Community strategy would then be perfectly defined by such programmes, and a strict application of the subsidiarity principle would mean that no additional proposal would be required on the part of the Community. Each nation would be responsible for putting its programme into effect and for its objectives being met. These objectives, when combined with one another, would lead to the overall result of stabilization.

An energy demand management approach
The factors described to this point with regard to the Spanish strategy for confronting the problem of climate change - National Climate Commission and National Programme for the limitation of CO_2 emissions - are important and are an adequate basis for dealing with this problem. But, in the long term, energy policy must be focused differently from the norm regarding National Energy Programmes. In energy regulation, managing energy supply is the usual focus, and increasingly greater emphasis muse be placed upon managing energy demand.

The concept of sustainable development, which was conclusively coined at the Rio Summit, implies the need to use natural resources and, in particular, energy resources, rationally. Managing demand means taking measures that will have an effect on energy consumption.

The main means that can be used as part of a rational energy demand management programme are as follows:

- **Economic instruments**. Within the framework of a market economy, material and cash-flows and the resulting cost-profit distributions are determined based on prices. Only by making the prices of energy resources gradually closer to their real value, as a result of internalizing the environmental cost in each one of the stages involved in obtaining, processing and using them, will it be possible for the market to function as a means towards the sustainable use of energy.

- **Information and education**. The measures taken in this area should be aimed at increasing average consumer awareness regarding all that has to do with the rational use of energy, through public service information campaigns, consultation and inspection procedures, energy-related labelling, etc.

It is equally important to improve the level of training of the pertinent groups of professionals in the energy-consumption cycle.

- **Regulatory measures**. Setting up energy-efficiency standards can be a useful measure for new buildings, equipment, vehicles and electrical appliances.

Experience reveals that to set up standards which will be effective, they must be ambitious but realistic, and must be set up in coordination with industry and other important groups, such as consumer associations.

- **Research and Development**. The importance of upgrading research efforts regarding energy technologies is obvious. Although this is particularly true of the energy production sector - more efficient power-production technologies - it is also of importance in the final consumption area: vehicles requiring less fuel, buildings requiring less power, etc.

One of the underlying ideas under the heading of energy demand management is that of the energy service concept instead of energy consumption. Energy in itself is of no direct value to man and, thus, it is wrong to consider energy consumption as an index of welfare. Energy is useful only as an input for processes that provide services: hot meals, transportation, lighting. These services provided using energy do, indeed, contribute directly to human welfare.

Full understanding of the fact that the true value lies in the service rendered by energy and not in energy itself, as well as making a gradual distinction between processes normally considered to go hand in hand, like energy consumption and economic growth, are fundamental requisites for sustainable energy planning, which should be the object of our efforts in the near future.

ssions policy and industry

npson
Execuⁱᵛᵉ Director
Imperial Chemical Industries PLC
London

Introduction

CO_2 is the principal greenhouse gas and most of this CO_2 comes from energy production. The economical provision of energy is essential to economic growth, and economic growth is essential to provide the resources for environmental improvement. Thus, we have a problem. Unless we can purchase energy economically with no, or at least less CO_2 emissions, or produce economic growth with less energy, we cannot achieve the optimum rate of sustainable development, both economically and environmentally, that we are seeking.

One solution is to develop new sources of energy which do not create CO_2. Without going into this in detail my assessment is that this is difficult, costly, and will definitely take a long time to have an impact.

I conclude, therefore, that while new sources of energy without the by-product of CO_2 are a possibility for the future, they will not provide a short-term answer. We must, therefore, seek either improved efficiency in producing energy from existing fuel sources or economic growth with less energy input. It is this latter area, energy efficiency, particularly in industry, on which I would now like to dwell.

Energy use and industrial performance in the UK

I would like to start by looking at the pattern of energy use in the UK. The three key sections are industry, transport and domestic use; and each accounts for about 30% of the nation's energy use. The chemical industry in which I work uses about one-fifth of the total industry consumption of energy, or 6% of the UK total. ICI itself accounts for about 10% of the UK industrial energy demand. Our consumption of electricity alone is 1% of the UK total.

When we look at carbon dioxide emissions in the UK, the picture is similar, although influenced by the source of energy. Thus, roughly one third of the UK's total emissions of some 600 million tonnes of CO_2 come from industry, if we include that generated in producing electricity used by industry.

But, if we now look at how industry has performed in reducing energy usage, the record is quite good. On a global basis, industrial energy efficiency has improved significantly. An examination of the energy intensity of the UK economy makes the same point, as most of this improvement has come from the industrial sector.

This performance by industry reflects the fact that energy costs have a direct bearing on bottom-line profitability. In the UK, for example, manufacturing industry now uses 25% less energy than it did in 1970, and the proportion of energy it used dropped from 42% in 1960 to 28% in 1989. This is in marked contrast to the continuing rise in energy used in transportation, and a slight increase in the domestic sector. This is reflected in carbon dioxide emissions from the various sectors.

The good energy efficiency performance of industry reflects two factors. The first is the changing mix in UK industrial output, a greater reliance on products such as pharmaceuticals and electronics and on service industries, and less on energy-intensive products such as steel or petrochemicals. This is both good and bad news depending on your perspective. There is no doubt, for example, that high electricity costs in the UK are tending to drive some segments of industry abroad. But, in addition to the mix change, a lot of work has been done, particularly in the energy-intensive industries where the economic incentives and paybacks are the greatest.

However, for industry generally, energy costs are a relatively small part of their total cost structure, generally less than 3%. In these times, when every business is watching its costs very carefully, it is difficult to get CEOs to give the necessary priority to energy conservation activities. When capital is under constraint, investment in energy-saving projects requires the attention of top management if it is to receive sufficient priority to be implemented.

The ICI experience
Recognizing the need for encouragement, it may be worth a few minutes just to look at the experience of one energy intensive industry - the chemical industry, in which ICI participates.

The UK chemical industry is very energy intensive, using about one-third of all energy utilized in industry, if feedstocks are included, compared with its 10% share of total UK manufacturing. The chemical industry has made continual improvements since the 1960s.

Even from 1980 to 1987, UK chemicals output grew by 32% while energy purchases continued to be reduced by 15%. This represents a 36% decrease of energy use per unit of output, and has been achieved by new technology and new process equipment, as well as better energy management and more efficient plant operations.

Specifically, in ICI worldwide since 1971, that is in the last 20 years, production has more than doubled while energy use has actually been reduced by 10% and CO_2 emissions by 20%. This means that we have more than halved energy use per unit output while reducing CO_2 per unit of output by two-thirds.

Although this may seem impressive, we believe there is still much to be done. In fact, one of the key environmental objectives established by the Chairman of ICI, as announced last year, was to implement an even more rigorous energy and resource conservation programme, from which we expect to have achieved substantial benefits by 1995.

In considering which additional steps need to be taken in the future, it is interesting to see how we have made progress in the past.

Undoubtedly, the two oil shocks in 1973 and again in 1979 played a big part. The almost instantaneous leap in oil prices and the magnitude of these changes forced all companies to look at energy costs and to find ways to reduce them. This was particularly true in the energy intensive industries. This might lead one to conclude that a further rapid rise in oil prices is necessary to galvanize further action; and it is this line of thinking that has led to consideration of a carbon tax. However, as I hope to show later on, a carbon tax is a very blunt instrument and doesn't seem, to most people in industry, to be the most efficient or effective way to promote greater energy efficiency.

But what the industry has demonstrated is that a great deal can be accomplished by attention to detail and the use of well-known procedures and equipment; by proper maintenance of steam traps, proper insulation of pipes and equipment, and attention to effective operation of energy production and energy using equipment. There is also, for example, increasing use of computerized systems of energy management. In general, our technical and engineering people believe that these sorts of measures, conscientiously applied, can result in energy savings of at least 10% in most circumstances.

Industrial potential and encouragement

This was confirmed by a UK government study[24] which estimated that for industry generally, good housekeeping and better energy management would produce savings of 10%. Retrofitting improvements to existing equipment, where the investment had a pay-back of less than 2 years, would produce a further 10% saving; and if the pay-back were stretched to 5 years, a further 10% was available. This total potential of 30% is well worth going for, and the key is to get industry motivated.

The Government's Energy Efficiency Office, now part of the Department of the Environment, is seeking to gain the commitment of top management across the range of industrial users by its 'Making a Corporate Commitment' campaign. This seeks to enrol companies in a commitment to adopt energy efficiency objectives, develop plans to achieve them, maintain and report on progress and educate employees so they can play their full part. This got off to a successful start last year and is receiving increased emphasis at a series of seminars chaired by Government Ministers over the next few months. The campaign basically is designed to get CEOs committed to energy efficiency.

---

24. Energy Paper 58, *An Evaluation of Energy-related Greenhouse Gas Emissions and Measures to Ameliorate Them* (UK country study for IPCC), HMSO, January 1990.

For many small and medium sized companies, and possibly for some large ones as well, even where the top level commitment is there, what is needed is guidance on practical steps. In recognition of this fact, the Advisory Committee on Business and the Environment, a committee of businessmen and women established by the Secretaries of State for the Environment and the DTI, has recently co-operated with the Energy Efficiency Office to produce a simple work book, step-by-step guide to actions that can be taken. This guide will be published in November[25] and I hope widely disseminated through regional workshops and meetings led by a variety of bodies such as Business in the Environment, the CBI, the Environmental Management Group of the ACBE, the DTI and so on.

In addition to good energy management, there are other steps which concerned managers can take. These include considering the efficiency of existing power and steam plants and also looking at how technology can improve the energy efficiency of the process.

Specific steps to recover waste heat may require investment but can produce quite significant energy savings at acceptable rewards. The choice of fuels, and particularly today's lower reliance on coal, has had an effect on CO_2 emissions.

There are, of course, many ways to improve the energy efficiency and fuel conservation of existing boiler steam plants. Modern conventional power stations convert only about 36% of the fuel into electricity, with most of the rest wasted as low grade heat. If both heat and power are used, the efficiency is much improved. Combined heat and power plants (CHP) are particularly suitable for industrial applications, as energy efficiency can be doubled in comparison with conventional power stations, reaching energy efficiencies in the range of 85%.

Through a series of process design and technology changes, the energy consumption per tonne of ammonia produced has been reduced by more than two-thirds by the change from the old coke-based process to today's modern plants. The latest process design, for which ICI has won an environmental award, our LCA process, has brought about reductions of 87% of NO_x, 95% of SO_x, 60% of CO_2 and 75% of ammonia in liquid effluent in comparison with the previous standards. If all ammonia world-wide were produced in this way then, in terms of nitrous oxide emissions, it would be equivalent to taking more than five million cars off the road. I am sure that almost every industry can find similar technology improvements if they are systematically sought out.

Industry will undoubtedly also have a role to play in improving energy efficiency in the non-industrial sectors as well.

---

25. *The Practical Energy Saving Guide*, Department of Energy/ Energy Efficiency Office, November 1992.

Transportation

I would like to consider first the transportation field which accounts for a growing proportion of CO_2 emissions. Private and business motoring accounts for about two-thirds of the total CO_2 emissions in the transport sector, or about 13% of total UK CO_2 emissions. It is difficult to see how the UK Government, or indeed the global community, will achieve its target of stabilizing or reducing CO_2 emissions without achieving some progress in this field.

One solution would be alternative fuels, but even if for example we could solve the problems and switch to electric cars, there remains the problem of how this electricity is generated. Pending some breakthrough in technology, for example cars fuelled by hydrogen, it would seem the answer lies in either reducing automobile usage, or improving automobile engine efficiency.

I don't think it is possible to reduce automobile usage by directive. The trend in the world is to greater human mobility.

In the circumstances, it would seem the best bet is to increase automobile engine efficiency. This is an area where the Government might well use economic instruments to add to the pressure of market forces in order to bring about change. An example could be to make engine efficiency a factor in determining the tax treatment of company cars.

But within the existing system there is a lot that companies can do, particularly here in the UK to help make progress. In the UK some 55% of new car registrations are company cars. Thus companies have it in their power to influence both automobile design and the way in which these cars are used. For example, companies can take into account engine efficiency in deciding what cars to purchase. The Department of Transport in the UK produces an annual rating of all cars sold in this country based on fuel efficiency. It is in the interest of the company to buy fuel-efficient cars. If fuel efficiency became an important parameter in directing company car purchases, I am sure we could see further dramatic improvements. And this does not mean just purchasing smaller cars. Depending on usage a 2.0 litre engine can show better fuel efficiency than a 1.5 litre one - and vice versa.

Domestic energy use

I have not so far addressed the third area of energy use - that is domestic or home use. Let me touch briefly on two areas.

The first is lighting. It is estimated that energy savings of up to 30% are possible with the use of existing low-energy lighting. However, the likelihood of achieving this under current conditions is problematical. The typical payback for home-owners is 0.7-2.0 years, depending on the load factor. Also, such bulbs are hard to find and expensive to purchase. A good place to start would certainly be in industrial, ie office, use, and in commercial establishments. In a very interesting report on 'Energy Efficiency in UK Supermarkets', Sainsburys has shown how they have cut energy consumption per unit of lighting to one third of what it was in the 1960s, with a great deal of this improvement in the last few years.

This is an area where industry can take the lead by ensuring that it is using efficient lighting - another area where good environmental practice and cost savings go hand in hand.

The second area is home insulation. The development of one clearly defined home energy rating scheme would help. Here the Government has done this in one form by devising a system that enables the two current schemes, the National Home Energy Rating Scheme (NHERS) and the Star Point system, to be related. With one clear labelling scheme, it would be possible to require that all new houses meet a certain standard, and then extend this to require energy rating of all existing houses, particularly when a house is sold. I believe this would motivate consumers to improve their homes. Current building regulations are Grade 6 but are due to be upgraded in 1993. The potential for improvement is obvious. Industry should welcome such a move. At the present time capacity exists to insulate only about 75,000 to 100,000 homes per year. The housing stock is something like 20 million homes. If the market develops it will provide significant business opportunities and employment.

Economic instruments and the carbon tax
The point of looking at these various steps is to stress that industry is very effective at marshalling sources, developing new products, and using its ingenuity to bring about changes usually more quickly than even it predicts. But the driving force to do this is competition and market forces. If the Government is to intervene to bring about more rapid changes than would otherwise occur it needs to do so in a way that reinforces this market-driven, competitive pressure.

There is now much talk about carbon taxes or carbon/energy taxes and the role they might play in achieving these results. While as you can tell from what I have said today that I am not opposed to the use of economic instruments to help achieve agreed national objectives, I am opposed to the carbon tax as it is currently proposed.

Broadly speaking, as I am sure you all know the proposal is for a carbon tax starting at the equivalent of $3 a barrel and using to $10/barrel of oil equivalent to be applied 50% on carbon content and 50% on energy content. This relates to the current price of oil of about $20 per barrel. However, the carbon tax is a very blunt instrument and in my view, will not only be costly, it will be ineffective. The tax is supposed by to be fiscally neutral but this will be almost impossible to monitor, and as different countries will use different means to achieve this fiscal neutrality (if they in fact do so) it will be distortive between countries.

In the industrial sector, because of its distorting effect, there will be various elaborate schemes to mitigate this effect. These schemes will be complex, bureaucratic, and expensive to monitor. The job creation in the government sector will be enormous but is this really what we want?

In the transport field its effect will be negligible. The impact on petrol prices will be around 6% at the pump. In the last two weeks we have seen changes of this magnitude or greater because of the change in exchange rates and certainly fluctuations in the normal course of business are common. So it will not be effective there.

And in the domestic sector, the problem is motivating home owners to take action and it is clear that much more targeted incentives are necessary for that to happen. As one example there is now a 17.5% VAT on insulation materials, but no VAT on energy consumption. Is this giving the right signal? A change here, if a signal were necessary, would be more effective than a blunt carbon/energy tax.

I believe more specially tailored economic instruments could be effective, and during this paper I have suggested a few. I hope the Government listens to the arguments of the CBI, and other industrial bodies, as well as private industry, and does not support an immediate move to a carbon tax - certainly not until the case is more proven and there has been further study of alternatives. If it is concluded that some intervention is required, I think it is fair to say that there are a number of 'no regret' policies that could be adopted and which would be effective before such a potentially costly intervention as a carbon tax is imposed.

Energy efficiency policy in the European Community: how CO_2 emissions may (or may not) be reduced

Andrew Warren
Director
Association for the Conservation of Energy
London

Between 1974-84, across the European Community, energy efficiency improved by an average of 20%. This meant producing more wealth for the same or even smaller amounts of fuels consumed. It reduced the number of new power stations and gas and oil fields required. Over the same period, however, Japan's efficiency improved by 34%.

Now a further 20% savings target - set by the Community between 1985-1995 - is being allowed to slip. Investments in improving energy performance have declined - in some countries dramatically. Between 1985 and 1990 a meagre 7% improvement overall was achieved. The European Commission has noted:

'If current trends in the consumption and efficient use of energy continue, there is little hope of the Community achieving its 1995 objective of improving by 20% the efficiency of final demand. Failure to achieve this will have serious consequences for energy supply, European competitiveness ... and the environment'.

The SAVE programme (Specific Actions for Vigorous Energy Efficiency) was the attempt by the Commission to reverse this trend.

In 1990, the European Community emitted 2738 million tonnes of carbon dioxide. When stabilization by 2000 was originally agreed in October 1990, the anticipated business-as-usual increase was 11%.

Simultaneously, East Germany joined the Community. Incorporating it, and acknowledging 4% growth in CO_2 emissions in the old Community during 1991 alone, the EC now estimates an increase of 14% in CO_2 levels by 2000 (based upon 2.4% GDP growth). Because of its acknowledged inefficiency, the EC now assumes that East Germany's CO_2 emissions will decrease substantially (by 20 million tonnes over the decade), thus bringing the anticipated EC growth back to 12%.

However, SAVE's original conception dates back further. It goes back to a paper produced on May 13 1987, entitled: 'Towards a Continuing Policy for Energy Efficiency in the European Community' (COM(87) 223). From this was developed the new programme entitled SAVE. Nobody should doubt the considerable expectations with which this programme was conceived.

For instance, in a proposal for a Council Decision issued by the Commission on May 12 1989, in paragraph 15 it is stated quite specifically that 'a realistic target for SAVE would be to maintain energy consumption in 2010 at the 1988 level'.

At the Council meetings on May 31 and October 1 1990, the Council of Ministers expressed favourable opinions about the SAVE programme, and called for it to be formally adopted as soon as the opinion of the European Parliament had been received.

On November 13 1990, a 'proposal for a Council decision concerning the promotion of energy efficiency in the Community' (COM(90) 365 final) was sent to the President of the Council. This document set out in considerable detail all the proposed directives under the SAVE programme. They fell under three different headings: technical measures; financial and taxation measures; and measures relating to user behaviour.

Under technical measures, particular concentration was placed upon the building sector, which uses over 40% of final energy consumption - an irony bearing in mind the subsequent subsidiarity arguments dredged up to kill off SAVE.

In Annex 1 of that document, which was criticized by the European Parliament for having insufficient finance and ambition, a chronology of legal actions and standards was included. This detailed very clearly some 13 different areas to be covered in each half year, beginning in the second half of 1990 with building certification, heat generator standards, and heat metering on the basis of actual consumption; and closing in the second half of 1992 with motor vehicle performance requirements and minimum standards for a range of domestic appliances not previously covered.

Subsequently, on various occasions, a variety of different members of the Parliament asked the Commission about the progress - or rather lack of it - of many of these promised legal actions.

Meanwhile time moved on. On October 16 1991, the Commission formally adopted its key strategy paper for the community to limit carbon dioxide emissions and improve energy efficiency. In practice, versions of this paper had been widely circulated for much of the previous six months.

According to Annex 6 of this document, the total additional savings of CO_2 from SAVE would be 77.8 or 15% of the 526.4Mtoe of CO_2 required to be reduced. However, as a footnote, it was added that 'it has to be noted that the impact of some SAVE measures are already included in gains from the market and other policies'.

In paragraph 14 of this key document, it stated quite specifically that 'a set of regulatory measures will need to be developed...many of these are covered to some extent by Commission proposals like the SAVE programme, [which] will need to be strengthened'.

Within these sectoral measures, there were due to be sections on power generation, including a promise for a new proposal on least-cost planning; industry; transport; and household/commercial. On all of these, further regulatory measures were promised.

The final statement of the original SAVE programme was prepared under the Council decision 91/565/EEC and was placed in the official journal on January 30 1992 (9YC23/04). This provided a short description of the SAVE programme, as originally proposed by the Commission in its document (COM (90) 365) to which I referred earlier, which was amended by the European Parliament at its session in July 1991.

The clock ticked on. Early in 1992, the Rio summit was looming. The Commission wanted to have a substantial presence at the conference, to demonstrate that it was leading the field. The 'carbon strategy' announced the previous year had inevitably focused on the novelty of the proposed new carbon/energy tax. This was to be the Big Idea the EC would take to Rio.

To its proponents the tax had become a kind of macho-symbol for the true environmental believer. Other parts of the Commission were less convinced. But in a bargaining session in meetings of Chefs de Cabinets in late April and early May, a deal was struck. The tax would be endorsed - with the familiar caveats of 'conditionality' and 'opt-outs'. But as a quid pro quo, the related programmes, SAVE and its sister for renewable energy, ALTENER, were effectively emasculated, under the convenient guise of subsidiarity.

In practice, this was far more an example of good old-fashioned power politics horse-trading. Which did the environment part of the Commission want most - a series of detailed, technical, almost mundane initiatives, which individually might seem pedestrian, or a Big Idea? No matter that on their own figures SAVE could contribute every bit as much as the tax could (3% each, of the projected emissions). SAVE was duly sold down the river. The concept of a super SAVE programme, championed only months before, was conveniently forgotten.

Instead of issuing a series of specific directives on different topics, as originally planned, SAVE has now been redesigned as a single draft directive which outlines the advantages of pursuing action in seven areas:

- Energy certification of buildings
- Heat metering of buildings
- Insulation standards
- Boiler inspections
- Car efficiency inspections
- Third party financing in the public sector
- Company energy audits

Because the first three of these cover buildings - a non-tradeable good - the European Commission is particularly aware of the 'subsidiarity' issue, i.e. that such measures should be taken at the level of Member States. In fact the draft Directive takes this to such a degree that it does not specify what Member States should pursue in any of these areas in any detail - the actual standards are left up to them.

So now what do we have? Instead of a series of binding Directives, the European Parliament now has before it a paltry document. I gather now even the legality of this is being challenged by the Council, who say that the Directive should not be advanced under the environment Article 130S, under which it could be implemented with a qualified majority vote; instead it is claimed that it should proceed under the catch-all portmanteau Article 235, in which case it will require unanimous approval in Council.

But claims are still being made that SAVE will save 3% of emissions and that it will cut 61 million tonnes (rather than 77.8 million tonnes as before) - which will apparently be 'primarily achieved' by this paltry proposal. But nowhere does it specify what Member States will do in detail. The actual standards, even the timing are left up to them.

Let us take two examples, for both of which I have in my possession the original drafts which the Commission had prepared. They bear no resemblance whatsoever to what is now being placed upon the table. Let us take third party financing. To satisfy Article 4, Member States are merely asked to take the necessary measures to 'favour third party financing for investment'. What are those necessary measures? Will they be satisfied merely by issuing a few leaflets? Certainly, there is nothing binding whatsoever within this Article regarding specific actions.

Then let us take energy certification in buildings. Member States are merely asked to take appropriate measures 'in order to progressively' bring such certification into effect. By when? No firm timetable is suggested.

Under Article 189 of the Treaty, it states quite specifically that 'a directive shall be binding as to results to be achieved upon each Member State to which it is addressed'. I submit that the Parliament's legal services should consider: is this a Directive? Or is it just a Recommendation? There are no target results specified, certainly no quantifiable ones. It is a completely ill-defined document.

After two years, Member States are supposed to report upon progress. In theory, the Commission could then prosecute under Article 169 for lack of activity. But how are they to argue that there has been non-compliance if there have been no targets set and no dates?

This is truly a test case for 'subsidiarity', to see whether Member States are genuinely delivering their part of the bargain. The auguries are not good. Back in 1990, every Member State said it would provide the Commission with details of the CO_2 abatement plans. At subsequent Council meetings, it has been agreed over and over again that all Member States should try to do so. But even now, not every Member State has yet provided details even of their aspirations, let alone their detailed plans.

How can we know how the Community is doing, in working towards its agreed carbon targets, if we do not know how its component parts are faring? No wonder the long-promised monitoring mechanism is desperately needed.

Energy conservation is the cheapest, swiftest, most effective means of achieving targets. Such a statement has become almost a mantra for everyone examining this issue. It is agreed that to achieve our objectives we need a regular 2.5% per annum reduction in overall energy intensity. At the moment, we appear to be going backwards - in many places we are actually becoming more inefficient by the year. And yet the potential is so great, as the original SAVE programme set out so lucidly: 'a realistic target for SAVE would be to maintain energy consumption in 2010 at the 1988 level'.

Distributional effects and the use of the revenues from a carbon tax

Stephen Smith
Deputy Director
The Institute for Fiscal Studies
London

and
Jean Monnet
Senior Lecturer in European Economics
University College London

A carbon tax on the scale proposed by the EC would raise substantial revenues. These offer the potential for tax reforms which reduce the overall distortionary cost of the tax system by allowing the revenues raised from other, more distortionary, taxes to be reduced (the 'double dividend' argument). Alternatively, some, or all, of the revenues may be used to offset some of the distributional consequences and other side effects of the tax:

- A carbon tax would have a regressive[26] impact on the household income distribution in some EC countries, where substantial amounts are spent by households on domestic heating. This distributional effect can be offset by some form of lump-sum return of revenues to households; this will be more than enough to fully compensate poorer households. A package of measures which would achieve this lump-sum effect could include increases in income tax allowances, and increases in the level of pensions and state social security benefits.

- Concerns have arisen about the impact of a carbon tax on industrial costs and competitiveness. The revenues could be used to make adjustments in other cost items that would, on average, compensate for the higher taxes (although under-compensating energy-intensive sectors, and over-compensating those using little energy). Measures which might have effects on industrial costs would include reductions in payroll taxes, taxes on business assets or on profits. One way in which these tax reductions might be delivered is in the form of 'tax expenditures' (increased investment allowances, accelerated depreciation, etc).

---

26. By 'regressive' it is meant that the extra tax would constitute a higher proportion of the incomes of poorer households than of richer households.

- Price effects. It is not obvious that a revenue-neutral carbon tax should have any impact on the rate of inflation; for it to do so, there must be some asymmetry in responses to tax increases on carbon-based energy and tax reductions elsewhere. If inflation effects are feared, reductions in indirect taxes like VAT, sales taxes, and excise duties can be made to offset the impact of carbon taxes on the overall price level.

- The revenues could also be used to finance increases in certain public expenditures - for example, R&D subsidies or 'tax expenditures' to encourage greater investment in more energy-efficient equipment. There is an issue about how far the linkage between revenues and spending should be formalized. There may be some political attractions in earmarking carbon tax revenues to 'good' environmental purposes rather to general public spending. But earmarking can lead to long-run inefficiency in public spending decisions. Nevertheless explicit 'packaging' of a carbon tax with a statement of how the revenues are to be used seems likely to help secure political acceptance. The example of Sweden, where the carbon tax was presented in a package including explicit cuts of other taxes, may be a useful practical model for other countries to follow.

EC implementation: options and constraints

Nigel Haigh
Institute for European Environmental Policy
London

1. Relative roles of EC and Member States

I must quarrel with a sentence in the organizers' agenda for this meeting which
states: 'The debate on stabilizing CO_2 emissions has reverted to national policy
makers.' Surely it never left the hands of the national policy makers? The EC is
largely a legislative body or provider of funds. Practical action to curb CO_2 has
to be taken in the Member States. Even the carbon tax will be a national tax.
From the moment global warming reached the EC agenda the question has been:
how will policy be shared between the EC and Member States? It has therefore
involved the tricky subjects of 'competence' and 'subsidiarity'.

2. The EC stabilization initiative of October 1990

This was stated as a political objective by a joint Energy and Environment
Council - but in a qualified form and without legal force. It played an important
role in creating the political climate that culminated in the Climate Change
Convention.

3. The loss of momentum

Why then did it take eighteen months for the Commission to translate the
political declaration of October 1990 into the proposal for a Council Decision?
The proposal - COM(92)181 - is known as the 'monitoring mechanism' but as
presently formulated is in fact much more than that. By formalizing the political
objective in a legal text it becomes the cornerstone of EC policy and also
applies pressure on the Member States to take action themselves.[27] However,
the proposal is deficient in not requiring national programmes to be precise
enough, or to be published.

What happens if the national contributions do not add up is a very good
question.

4. Constraints on EC policy: 'competence' and 'subsidiarity'

'Competence' and 'subsidiarity' in relation to global warming are difficult
subjects, but will deeply influence what the Member States allow the EC to do.
The importance of 'competence' in relation to global warming was underlined

---

27. See the Postscript to this report.

62

by the Council when it agreed to sign the Convention. The subject has been discussed by me elsewhere.[28]

'Subsidiarity' in relation to global warming raises not just the question of what value EC policy adds to that of the individual Member States, but what EC policy adds to that embodied by a 'higher' level of 'government' eg. the machinery established under the Convention. The EC certainly has a role to play, but it is a different role from what it would have been in the absence of a convention

5. Options: 'integration'
As well as proposing EC policy on global warming by placing obligations on the Member States, and ensuring that jointly the Member States fulfil their convention obligations, the Commission also has the task of ensuring that the policies of all DGs (energy, transport, agriculture, internal market, etc) take account of the need to curb greenhouse gases. This is a Treaty obligation: 'Environment protection requirements shall be a component of the Community's other policies' (Article 130R).[29]

28. Nigel Haigh, 'The EC and International Environment Policy', in A. Hurrell and B. Kingsbury (eds), *The International Politics of the Environment,* OUP, 1992. This appeared earlier in *International Environmental Affairs* Vol.3, No.3, 1991.

29. David Baldock et al, *The Integration of Environmental Protection Requirements into the Definition and Implementation of Other EC Policies,* Institute for European Environmental Policy, London, 1992.

Postscript:
Results of the Environment Council meeting on 23 March 1993

Final production of this report was delayed until after the outcome of the EC Environment Ministers' Council on 22-23 March 1993, which considered EC policy relating to CO_2 emissions and the UN Framework Convention on Climate Change.

This meeting adopted as a Council Decision a modified version of the 'monitoring proposal', and reached political agreement also on the basis for Community ratification of the Convention. However, at that Council, six Member States declared that they believed that the stabilization of CO_2 emissions in the Community as a whole was not possible unless Community measures including a carbon/energy tax were adopted, and that if decisions on these were not taken in time the ratification agreement would have to be reconsidered.

Compromise text for the Monitoring Decision was presented by the Presidency on 9 March 1993. This was significantly modified from the earlier drafts available for discussion at the RIIA Workshop. The Monitoring Decision establishes the legal basis for future EC climate policy and contains key references to the Climate Convention. Because of its central importance in forming the context for future development of EC climate policy we here reproduce key extracts of the text (amendments made at the Council meeting, said to be minor, were not available at the time of going to press).

Proposal for a Council Decision for a monitoring mechanism of Community CO_2 and other greenhouse gas emissions
9 March 1993

THE COUNCIL OF THE EUROPEAN COMMUNITIES,[30]

Having regard to the Treaty establishing the European Economic Community, and in particular Article 130s thereof [which states that the EC shall maintain a high level of environmental protection] ...

---

30. Only a small part of the Preamble and selected Articles are included. Omitted Articles are numbers 1 (Declaration); 3, 'Inventories and data reporting'; 4, 'Procedures and methods for evaluation'; 7, 'Other greenhouse gases'; and 8, 'Committee'.

Whereas the Council of Environment and Energy Ministers agreed at their meeting on 29 October 1990 that the Community and Member States, assuming that other leading countries undertook similar commitments, and acknowledging the targets identified by a number of Member States for stabilising or reducing emissions by different dates, were willing to take actions aimed at reaching stabilisation of the total CO_2 emissions by 2000 at the 1990 level in the Community as a whole; and that Member States which start from relatively low levels of energy consumption and therefore low emissions measured on a per capita or other appropriate basis are entitled to have CO_2 targets and/or strategies corresponding to their economic and social development, while improving the energy efficiency of their economic activities;

Whereas the Council of Energy and Environment Ministers at their meeting on 13 December 1991 invited the Commission to propose concrete measures arising from the community strategy and required that such measures should take into account the concept of equitable burden sharing, according to the conclusions of the 29 October 1990 Council; ...

Whereas on the occasion of the signing of the Convention the Community and its Member States reaffirmed the objective of stabilisation of CO_2 emissions by 2000 at 1990 level in the Community as a whole, as referred to in the Council conclusions of 29 October 1990, 13 December 1991, 5 May 1992 and 26 May 1992; ...

HAS ADOPTED THIS DECISION:

Article 2
National programmes

1. The Member States shall devise, publish, and implement national programmes for limiting their anthropogenic emissions of CO_2 in order to contribute to:

- the stabilisation of CO_2 emissions by 2000 at 1990 levels in the Community as a whole, assuming that other leading countries undertake commitments along similar lines, and on the understanding that Member States which start from relatively low levels of energy consumption and therefore low emissions measured on a per capita or other appropriate basis are entitled to have CO_2 targets and/or strategies corresponding to their economic and social development, while improving the energy efficiency of their economic activities, as agreed at the Energy/Environment Councils of October 1990 and December 1991, and

- the fulfilment of the commitment relating to the limitation of CO_2 emissions in the UN Framework Convention on Climate Change in the Community as a whole.

These programmes shall be periodically updated.

2. Each Member State shall, at the latest from the first updating, include in its national programme:

- its 1990 base year anthropogenic emissions of CO_2 [ ] determined in accordance with the provisions of Article 3, paragraph 1;

- inventories of its anthropogenic CO_2 emissions by sources and removal by sinks determined in accordance with the provisions of Article 3, paragraph 1;

- details of national policies and measures which contribute to the limitation of CO_2 emissions;

- trajectories for its national CO_2 emissions between 1994 and 2000;

- measures being taken or envisaged for the implementation of relevant Community legislation and policies;

- a description of policies and measures in order to increase the sequestration of CO_2 emissions;

- an assessment of the economic impact of the above measures. ...

Article 5
First evaluation of national programmes and of the
state of emissions in the Community

1. Member States shall forward to the Commission their existing national programmes one month after the notification of this decision to the Member States.

2. The Commission shall forward to the other Member States the national programmes received within two months of their reception.

3. The Commission shall evaluate the national programmes, in order to assess whether progress in the community as a whole is sufficient to ensure fulfilment of the commitments set out in Article 2(1).

4. The Commission shall report to the Council and the European Parliament the results of its evaluation within six months of the reception of the national programmes.

Article 6
Subsequent evaluation of progress

After the first evaluation referred to in Article 5, the Commission shall annually assess in consultation with the Member States whether progress in the Community as a whole is sufficient to ensure that the Community is on trajectory to fulfil the commitments set out in Article 2(1) and report to the Council and the European Parliament, on the basis of information received under Articles 2 and 3, including where appropriate the updated national programmes.